T0254751

Lecture Notes in Computer Science 10686

FoLLI Publications on Logic, Language and Information

Subline of Lectures Notes in Computer Science

More information about this series at http://www.springer.com/series/7407

Annie Foret · Reinhard Muskens
Sylvain Pogodalla (Eds.)

Formal Grammar

22nd International Conference, FG 2017
Toulouse, France, July 22–23, 2017
Revised Selected Papers

Editors
Annie Foret
IRISA
University of Rennes 1
Rennes
France

Sylvain Pogodalla
LORIA/Inria Nancy
Villers-les-Nancy
France

Reinhard Muskens
Department of Philosophy
Tilburg University
Tilburg
The Netherlands

ISSN 0302-9743 ISSN 1611-3349 (electronic)
Lecture Notes in Computer Science
ISBN 978-3-662-56342-7 ISBN 978-3-662-56343-4 (eBook)
https://doi.org/10.1007/978-3-662-56343-4

Library of Congress Control Number: 2017962892

LNCS Sublibrary: SL1 – Theoretical Computer Science and General Issues

Printed on acid-free paper

This Springer imprint is published by Springer Nature
The registered company is Springer-Verlag GmbH, DE
The registered company address is: Heidelberger Platz 3, 14197 Berlin, Germany

Preface

The Formal Grammar conference series (FG) provides a forum for the presentation of new and original research on formal grammar, mathematical linguistics, and the application of formal and mathematical methods to the study of natural language. Themes of interest include, but are not limited to:

- Formal and computational phonology, morphology, syntax, semantics, and pragmatics
- Model-theoretic and proof-theoretic methods in linguistics
- Logical aspects of linguistic structure
- Constraint-based and resource-sensitive approaches to grammar
- Learnability of formal grammar
- Integration of stochastic and symbolic models of grammar
- Foundational, methodological, and architectural issues in grammar and linguistics
- Mathematical foundations of statistical approaches to linguistic analysis

Previous FG meetings were held in Barcelona (1995), Prague (1996), Aix-en-Provence (1997), Saarbrücken (1998), Utrecht (1999), Helsinki (2001), Trento (2002), Vienna (2003), Nancy (2004), Edinburgh (2005), Malaga (2006), Dublin (2007), Hamburg (2008), Bordeaux (2009), Copenhagen (2010), Ljubljana (2011), Opole (2012), Düsseldorf (2013), Tübingen (2014), Barcelona (2015), and Bolzano-Bozen (2016).

FG 2017, the 22nd conference on Formal Grammar, was held in Toulouse during July 22–23, 2017. The conference consisted in two invited talks, by Jakub Szymanik and Michael Benedikt, and nine contributed papers selected from 14 submissions. The present volume includes the contributed papers.

We would like to thank the people who made the 22nd FG conference possible: the invited speakers, the members of the Program Committee, and the organizers of ESSLLI 2017, with which the conference was colocated.

July 2017

Annie Foret
Reinhard Muskens
Sylvain Pogodalla

FG 2017 Organization

Program Committee

Berthold Crysmann	CNRS - LLF (UMR 7110), Paris-Diderot, France
Philippe de Groote	Inria Nancy – Grand Est, France
Nissim Francez	Technion - IIT, Israel
Thomas Graf	Stony Brook University, Stony Brook, USA
Laura Kallmeyer	Heinrich-Heine-Universität Düsseldorf, Germany
Makoto Kanazawa	National Institute of Informatics, Japan
Gregory Kobele	Universität Leipzig, Germany
Stepan Kuznetsov	Steklov Mathematical Institute, Moscow, Russian Federation
Robert Levine	Ohio State University, USA
Glyn Morrill	Universitat Politècnica de Catalunya, Spain
Stefan Müller	Freie Universität Berlin, Germany
Mark-Jan Nederhof	University of St Andrews, UK
Christian Retoré	Université de Montpellier and LIRMM-CNRS, France
Mehrnoosh Sadrzadeh	Queen Mary University of London, UK
Manfred Sailer	Goethe University Frankfurt, Germany
Edward Stabler	UCLA and Nuance Communications, USA
Jesse Tseng	CNRS, France
Oriol Valentín	Universitat Politècnica de Catalunya, France
Christian Wurm	Heinrich-Heine-Universität Düsseldorf, Germany
Ryo Yoshinaka	Tohoku University, Japan

Standing Committee

Annie Foret	IRISA, University of Rennes 1, France
Reinhard Muskens	Tilburg Center for Logic and Philosophy of Science, The Netherlands
Sylvain Pogodalla	Inria Nancy – Grand Est, France

Additional Reviewer

Mati Pentus	Moscow State University, Russian Federation

Abstracts of Invited Talks

Comparing Relational Vocabularies

Michael Benedikt

Department of Computer Science, Oxford University, Parks Road, Oxford, UK
michael.benedikt@comlab.ox.ac.uk
http://www.cs.ox.ac.uk/michael.benedikt/

A relational schema is a set of metadata describing relational instances, corresponding to tabular data. Schema information includes the names of relations, their arities, and optionally integrity constraints that capture some of the semantics of the data. In this talk I will outline research concerning the "expressiveness" of relational schemas. What does it mean to say that one schema subsumes the "information content" of another? How can one verify a schema-level relationship algorithmically? I will give a quick look at approaches to the problem proposed in database theory, including some work dating to the late 1970's and my own recent work in the area.

Hopefully the ideas proposed for comparing schemas for structured data can be of interest in comparing the expressiveness of vocabularies in other contexts.

From Grammar to Meaning

Jakub Szymanik

Institute for Logic, Language, and Computation, University of Amsterdam,
Amsterdam, The Netherlands
J.K.Szymanik@uva.nl
http://www.jakubszymanik.com/

I will discuss how different grammatical formalisms can be combined with logic and cognitive modeling techniques to account for the meaning of natural language. In a slogan, if you take care of the syntax of representational system, its semantics will take care of itself. I will survey some recent work on the meaning of quantifiers, reasoning, learnability, and evolution of language. The common thread of all the models will be taking the idea of logico-syntactic properties of thought (Language of Thought, LoT) seriously to account for linguistic and cognitive phenomena, showing how grammar can be driving the semantic engine.

The research leading to these results has received funding from the European Research Council under the European Union's Seventh Framework Programme (FP/2007–2013)/ERC Grant Agreement n. STG 716230 CoSaQ.

Contents

Binding Domains: Anaphoric and Pronominal Pronouns in Categorial Grammar

María Inés Corbalán(✉)

Universidade Estadual de Campinas, Campinas, São Paulo, Brazil
inescorbalan@yahoo.com.ar

Abstract. In this paper we present a treatment for anaphoric pronouns and reflexives in a Type Logical Grammar. To this end, we introduce structural modalities into the left pronominal rule of the categorial calculus with limited contraction **LLC** [8]. Following a proposal due to Hepple [6], we also sketch an analysis for the long-distance anaphora *seg* from Icelandic.

Keywords: 3SG Reflexives · 3SG Pronouns · Binding Theory
Type Logical Grammar · Calculus **LLC**

1 Introduction

From a generative perspective, the licensing of pronominal expressions such as *he*, *him*, *himself* is determined by the so-called Principles A and B of the Binding Theory [3]. Principle A stipulates that an anaphor (reflexives and reciprocals) must be bound in its governing category (roughly, it must have a c-commanding local antecedent). Principle B stipulates that a pronoun must be free (i.e. not bound) within its governing category; notwithstanding, a pronoun can be bound from outside this syntactic domain. Thus a pronoun, unlike an anaphor, also admits a free reading. Principles A and B jointly imply a strict complementary distribution between pronouns and reflexives in *some* syntactic domains, as exemplified below:

(1) John_1 admires himself_1/*him_1.

(2) John_1's father_2 loves $\text{him}_{1/*2}$/$\text{himself}_{2/*1}$.

(3) The father_1 of John_2 loves $\text{him}_{2/*1}$/$\text{himself}_{1/*2}$.

(4) John_1 believes himself_1/*he_1/*him_1 to love Mary.

(5) John_1 says $\text{he}_{1/2}$/*himself_1 loves Mary.

The Binding Theory has been successively revisited to overcome some counterexamples. Complementary distribution is disconfirmed, on the one hand, in adjunct clauses, like in (6) below. On the other hand, languages like Icelandic, Dutch, German and Norwegian each contain an anaphoric form—*sig, zich, sich,*

A. Foret et al. (Eds.): FG 2017, LNCS 10686, pp. 1–19, 2018.
https://doi.org/10.1007/978-3-662-56343-4_1

seg, respectively—that does not meet Principle A, as the former can be bound by a long-distance antecedent (cf. [22]).[1]

(6) John$_1$ glanced behind him$_1$/himself$_1$.

(7) Jón$_1$ segir að María$_2$ telji að Haraldur$_3$ vilji að Billi$_4$
John says that Maria believe.SBJ that Harold want.SBJ that Bill
heimsæki sig$_{1/2/3/4}$.
visit.SBJ SE-ANAPHOR
'John says that Maria believes that Harold wants that Bill visit him/her.'

Several categorial—combinatory and type-logical—calculi have been proposed to deal with reflexives and anaphoric pronouns. Some of them treat multiple-binding into the lexicon (cf. [16,23]); others use a syntactic approach (cf. [6–8]). Working on a Type-Logical Grammar, Jaeger [8] develops the Lambek calculus with Limited Contraction (**LLC**) to syntactically process personal pronouns in a uniform way; he does not discriminate syntactically (nor semantically) among reflexives, anaphoric pronouns and pronominals. In other words, he does not take Principles A and B of the Binding Theory into account.

Our goal is to give a more accurate treatment of personal pronouns, taking as a starting point the pronominal connective $|$ from Jaeger and the intuition behind its logical rules. Firstly, we modify the right rule of **LLC** to distinguish on the one hand the *free* (or pronominal) *use* from the *bound* (or anaphoric) *use* of a personal pronoun. Secondly, by using the (lexical) structural modality $\langle\ \rangle$ of [18] (and, analogously, for our $\lceil\ \rfloor$), we identify different syntactic domains for binding: we impose structural conditions into the pronominal left rule of **LLC** by using the corresponding (syntactic) structural modality $[\ ]$ (and also $\{\ \}$). Thus, on the other hand, we also distinguish anaphoric pronouns from reflexive anaphors. As a consequence, although we deal with reflexives, anaphoric pronouns and pronominals, our proposal is not intended to be a uniform approach. For reasons of space, we restrict ourselves to cases where the binder is a nominal phrase and the bindee (a pronoun or a reflexive) carries 3SG features.[2] Since our proposal is inspired by Jaeger's calculus, we also do not deal with cases in which the bindee precedes its binder.[3]

The structure of the paper is as follows. In Sect. 2, we present Jaeger's calculus **LLC** (in a sequent format) and we briefly discuss some questions related to the problem of overgeneration. In Sect. 3 we change the right pronominal rule of **LLC** to distinguish between a reflexive and a pronominal type-constructor. In Sect. 4, firstly we present our treatment for subject-oriented anaphors in several syntactic

[1] Although these languages contain this kind of simple (also weak) reflexive form, their syntactic behavior is not the same in all of them (cf. for example, [5]).

[2] Hence, we restrict ourselves to what some theories call anaphoric coreference, not binding (cf. [2,21]). Though it is generally accepted that reflexives and reciprocals behave in the same way with respect to binding conditions, their semantic value diverges. For this reason, we also do not deal with reciprocal anaphors.

[3] However, a version of Jaeger's rules that also allows cases of cataphora is presented in [17].

domains and secondly, we deal with object-oriented anaphors in double-object constructions and prepositional complements. Finally, we sketch an analysis for long-distance anaphors from Icelandic. Section 5 concludes the paper. In the Appendix we sketch the principal cut for our new pronominal rules.

2 LLC Calculus

LLC is a conservative extension of the Lambek **L** calculus (without empty antecedents) [9]. Like **L**, **LLC** is free of structural rules. Jaeger's calculus treats resource multiplication syntactically. **LLC** extends the sequent calculus **L** by adding the anaphoric type-constructor $|$. The rules of the latter encode a restricted version of the structural rule of Contraction, thus allowing for multiple-binding (see Fig. 1). Despite incorporating this structural rule, **LLC**, as well as Lambek system, enjoys Cut elimination, decidability and the subformula property. Indeed, as the reader can check, all the formulas that occur in the premises of the two new rules for the anaphoric type-constructor are subformulas of the formulas that occur in their conclusion.

$$\frac{Y \Rightarrow M : A \qquad X, x : A, Z, y : B, W \Rightarrow N : C}{X, Y, Z, z : B|A, W \Rightarrow N[M/x][(zM)/y] : C} \ |L$$

$$\frac{X, x_1 : B_1, Y_1, \ldots, x_n : B_n, Y_n \Rightarrow N : C}{X, y_1 : B_1|A, Y_1, \ldots, y_n : B_n|A, Y_n \Rightarrow \lambda z.N[(y_1 z)/x_1] \ldots [(y_n z)/x_n] : C|A} \ |R$$

Fig. 1. Left and right rules for $|$

Note that when A is a basic type, the left premise of $|L$ is an instance of the identity axiom; thus the rule can be simplified, as shown in Fig. 2.[4]

$$\frac{X, x : A, Z, y : B, W \Rightarrow M : C}{X, x : A, Z, z : B|A, W \Rightarrow M[zx/y] : C} \ |L$$

Fig. 2. Simplified left rule for $|$

Anaphoric expressions are then assigned a type $B|A$: it works as a type B in the presence of an antecedent of type A. The $|L$ rule expresses the fact that for an anaphoric expression to be bound it needs an *antecedent* in the same

[4] Jaeger is not only concerned with anaphoric pronouns but also with other anaphoric phenomena, such as ellipsis of VP.

premise, that is, in some *local* syntactic domain. Besides imposing an antecedent condition, this rule incorporates a restricted version of the structural rule of (long-distance) Contraction, in that the antecedent A for the anaphoric type $B|A$ occurs in both premises of this rule.

Since personal pronouns take their reference from a nominal antecedent, they are assigned the syntactic anaphoric category $n|n$.[5] In semantic terms, a pronoun denotes the identity function $\lambda x.x$ over individuals; the reference of a pronoun is identical with the reference of its antecedent.[6]

Since a pronominal type $n|n$ can be constructed by using the |R and |L rules, and since anaphoric and pronominal pronouns are assigned the same syntactic type (and the same semantic category), the system can accurately recognize the free and the bound readings for a pronominal. Thus, for example, the system recognizes the double reading for *he* in (5), and so may assign it the saturated type s or the unsaturated (or functional) type $s|n$. The latter expresses the fact that a free pronoun occurs in the clause. In addition, the system can also derive the co-occurrence of bound pronouns and reflexives in syntactic domains in which complementary distribution fails, as exemplified in (6) above (see Fig. 3). It can also recognize the ungrammaticality of (8b) below, since the antecedent condition on |L is not fulfilled. Nevertheless, **LLC** also allows for the ungrammatical anaphoric readings in the following examples.[7]

(8) a. John_1 saw himself_1/*him_1.
    b. * Himself_1 saw John_1.

(9) a. John talked to Mary_1 about herself_1.
    b. * John talked about Mary_1 to herself_1.

(10) John_1 saw *himself_1's/his_1 mother.

(11) a. John_1 believes himself_1 to kiss Mary.
    b. * John_1 believes himself_1 kisses Mary.

$$\frac{n, (n\backslash s)/pp, pp/n, n \Rightarrow s}{n, (n\backslash s)/pp, pp/n, n|n \Rightarrow s}\ |L \qquad \frac{n, (n\backslash s)/pp, pp/n, n \Rightarrow s}{n, (n\backslash s)/pp, pp/n, n|n \Rightarrow s|n}\ |R$$

Fig. 3. Schematic derivation for $John_1$ *glanced behind* $himself_1/him_1$

Since we are looking for a more accurate treatment for the distribution of pronominal and anaphoric pronouns, we shall begin by distinguishing between an

[5] As usual, we use n for proper names, s for sentences, cn for common nouns and pp for prepositional phrases.

[6] In this respect, Jaeger follows [6,7].

[7] Everaert [4] uses these sentences to evaluate the scope and limits of several generative models for binding.

anaphoric connective for reflexives and a (possibly) non-anaphoric connective for personal pronouns like *he* and *him*. Later on, we shall draw a distinction between reflexives and bound pronouns.

3 Bound and Free Pronouns: Splitting the Pronominal Connective

For non-reflexive pronouns, we adopt Jaeger's left rule and the following left and right rules, which split the |R rule of **LLC**. It is important to emphasize that these two new rules, like those of **LLC**, satisfy the subformula property: all the formulas that occurs in the premises of ||L and ||R are subformulas of the formulas that occurs in their conclusion. Given that the proof of Cut elimination for these rules requires using a limited version of the Expansion rule (see Appendix), we call our modified version of Jaeger's system **LLBE**: Lambek calculus with Bracketed Expansion (Fig. 4).

$$\frac{X, x : \llbracket A \rrbracket, Z, y : B, W \Rightarrow M : C}{X, Z, z : B \| A, W \Rightarrow M[zx/y] : C} \|L \qquad \frac{Y \Rightarrow N : C}{x : \llbracket A \rrbracket, Y \Rightarrow \lambda x.N : C \| A} \|R$$

Fig. 4. Left and right rules for ||

As can be noted, we split (an extremely simplified version of) the |R rule of **LLC** to obtain a second left rule.[8] Hence, a pronominal type-constructor will have two left rules: |L and ||L.[9] By breaking the |R rule of **LLC** we can more clearly show that free and bound are labels that result from the procedures by which we use a pronoun: we apply the rule of use |L to get a bound (or anaphoric) use of a pronoun, while in applying ||L we use a pronoun freely.[10]

The ||R rule compiles a restricted form of the structural rule of Expansion, as it introduces a formula that is a sub-formula of the pronominal type $C\|A$. Given that we do not assume logical rules for the brackets $\llbracket \rrbracket$, they can only be introduced (deleted) through the use of the ||R (||L) rule. Consequently, like in Jaeger's proposal, the rule of proof for a free pronoun goes hand in hand with its free use in **LLBE**. However, unlike the |R rule of **LLC**, the ||R rule of **LLBE**

[8] Strictly speaking, we split the |R rule of **LLC** for the case where $n = 1$ and X is the empty sequence ϵ. As we shall show in the Appendix, the proof of principal Cut for the new rules requires using bracketed versions of the structural rules of Permutation and Expansion. In order to avoid a proof of a pronominal type $C\|A$ for any type C, the antecedent type A of the rule ||R has to be left-peripheral.

[9] From Sect. 4, the |L rule (for non-reflexive pronouns) will be renamed $\|L_a$, and ||L will have to be read as $\|L_p$. Though we shall retain |L for reflexives only, we will rename it $|L_a$ for the sake of uniformity.

[10] As we shall see later, the formula B in the ||L rule will have a bracketed structure $[B]$ in most cases.

does not simultaneously construct a pronominal type to the left and to the right sides of a sequent.

In addition, since **LLBE** contains two left pronominal rules (i.e. $|L$ and $\|L$), we are able to characterize two anaphoric type-constructors: a reflexive type-constructor, which uses only the $|L$ rule, and a pronominal type, which uses $|L$, $\|R$ and also $\|L$.[11] By assigning different syntactic types for reflexives and pronouns—$n|n$ and $n\|n$, respectively—, and given that the $|R$ rule of **LLC** for the case $n = 1$ may be derived by using $\|L$ and $\|R$ of **LLBE**, the latter system, like the former one, adequately recognizes grammatical sentences like those in (12–16a), whilst blocking the ungrammatical sentence in (16b) below (Fig. 5).

(12) John$_1$ said he$_{1/2}$ runs.

(13) John$_1$ said Mary likes him$_{1/2}$.

(14) John$_1$ likes him$_2$.

(15) He$_1$ likes himself$_1$.

(16) a. John$_1$ likes himself$_1$.

b. * John$_1$ likes himself$_2$.

$$\dfrac{\dfrac{\vdots \atop n, (n\backslash s)/n, n \Rightarrow s}{[\![n]\!], n, (n\backslash s)/n, n \Rightarrow s\|n}\,\|R}{n, (n\backslash s)/n, n\|n \Rightarrow s\|n}\,\|L$$

Fig. 5. Derivation for *John$_1$ likes him$_2$*

Nevertheless, since $|L$ is adopted for the pronominal type $n\|n$ and also for the anaphoric type $n|n$, **LLBE** is not yet capable of separating bound (object) pronouns from reflexives.[12]

---

[11] A Type-Logical sequent calculus generally contains one left and one right rules for each type-constructor. Since in our proposal the reflexive type-constructor uses only a left rule, our approach is non-standard.

[12] In order to distinguish subject and object pronouns, we could assign the lifted type $(s\|n)/(n\backslash s)$ to the former (cf. [17]). Although at first glance it would seem that a lifted type—$(s/n)\backslash(s\|n)$—is also adequate to categorize an object pronoun like *him*, it is not clear how we could deal with Exceptional Case Marked (ECM) constructions, in which the semantic argument of the embedded infinitive clause surfaces with accusative case. Indeed, if *him* were assigned $(s/n)\backslash(s\|n)$ because of its surface form, it would combine with a verb phrase to the left, like a real object complement does. But if this were the case, the subject slot of the embedded complement clause would not be saturated and then, the sentential argument of the ECM verb would become unsaturated as well.

4 Reflexives and Bound Pronouns: Imposing Structure Through Bracket Modalities

4.1 Subject-Oriented Reflexives and Bound Pronouns

There are several syntactic domains where a reflexive can occur: in nominal object complements, prepositional object complements, adjunct clauses, and even in an embedded position within nominal phrases (the so-called NP anaphora). In some of them, complementary distribution is fully verified: when the reflexive and its antecedent are co-arguments of the same function, a bound pronoun is ruled out.

Propositional Complements. Complementary distribution is also verified in the opposite direction in (some) clauses selected by propositional verbs like *say* and *believe*. As is widely known, reflexives are ruled out and bound pronouns are licensed in (finite) propositional complements of both these verbs, as exemplified below:

(17) John$_1$ said/believes *himself$_1$/he$_{1/2}$ walks.

(18) John$_1$ said/believes Mary hates *himself$_1$/him$_{1/2}$.

Anaphors within propositional complements have already been adequately analyzed in Categorial Grammar (cf. [6,15,23], a. o.). The correct binding relation in these complements is ensured by using a normal (or semantic) $S4$ modality $\Box$ [13]. In these categorial proposals, reflexives and pronouns are assigned different pronominal types. The following modalized lexical entries capture the above-mentioned facts:

him/he : $\Box(\Box n \| n)$
himself : $\Box(n|n)$
say/believe : $\Box((n\backslash s)/\Box s)$
walk : $\Box(n\backslash s)$
hate : $\Box((n\backslash s)/n)$

Nevertheless, authors have sometimes glossed over the fact that a reflexive may occur within a propositional complement if it occupies an embedded position:

(19) Max$_1$ said (that) the queen invited both Lucie and himself$_1$/him$_1$ for tea.

Similarly, it has not always been noted that *believe* licenses the occurrence of a reflexive in the subject position when the complement verb is in a non-finite form.[13] This last fact is a specific case of a more general situation: in complements of Exceptional Case Marked (ECM) verbs, such as *believe* or *expect*, a reflexive is allowed, while a bound pronoun is ruled out in the subject (and also the

[13] In passing, we point out that, unlike English, literary Spanish and Italian allow a nominative free or bound pronoun in non-finite complements of propositional verbs [11].

object) argument slot. Thus, in ECM constructions, the claimed complementary distribution is verified, as in other verb complements. Nevertheless, complementary distribution in ECM constructions is, in some sense, unexpected, since the reflexive in the subject position of non-finite complements is not a co-argument of the binder.

(20) John_1 believes himself_1/*him_1/*he to kiss Mary.

(21) Lucie expects John_1 to like himself_1/*him_1.

(22) Lucie_1 expects herself_1/*her_1/*she to kiss John.

From this evidence, it seems important to differentiate the lexical entry for a propositional *believe* from the ECM *believe*, despite the fact that both verbs select a sentential (finite or non-finite) complement. Following [12,14,18], we shall use the structural modality $\langle\ \rangle$ to mark the (syntactic) argument positions of a verb; though we shall use it to mark not only the subject position, but also the object complement position. In order to distinguish the propositional verb *believe/say* from non-propositional verbs, we shall set a different bracket modality $\lceil\ \rfloor$ aside for the former.

A sample of a bracketed lexicon is given below:

walk : $\langle n \rangle \backslash s$
like/hate/kiss : $(\langle n \rangle \backslash s) / \langle n \rangle$
say/believe : $(\langle n \rangle \backslash s) / \lceil s \rfloor$
john/mary : n
him : $n \| n$
himself : $n | n$

The right rules for brackets $\langle\ \rangle$ and $\lceil\ \rfloor$ are given below. The rules for Lambek's slashes are applied to structured sequences $[X]$ and $\{X\}$ of types.[14] The structures $[\ ]$ and $\{\ \}$ are then spread over the sequents when functional types A/B and $B\backslash A$ are built out of [B] and $\{B\}$, respectively. Hence, while the structural modality $\langle\ \rangle$ is a lexical mark, the insertion of the modalities $[\ ]$ and $\{\ \}$, and thus the delimitation of syntactic domains, is a consequence of syntactic operations (Figs. 6 and 7).

$$\frac{X \Rightarrow A}{[X] \Rightarrow \langle A \rangle}\langle\ \rangle\mathrm{R} \qquad \frac{X \Rightarrow A \qquad Y[B] \Rightarrow C}{Y[B/A, X] \Rightarrow C}[/]\mathrm{L} \qquad \frac{X \Rightarrow A \qquad Y[B] \Rightarrow C}{Y[X, A\backslash B] \Rightarrow C}[\backslash]\mathrm{L}$$

Fig. 6. Rules for brackets $\langle\ \rangle$ and structured $[\ ]$ sequents

[14] Generally, $\Delta[\Gamma]$ indicates a configuration Δ containing a distinguished configuration Γ of types. In our rules, $X[Z]$ would indicate a *sequence* X with a distinguished structured sequence $[Z]$ of types, and analogously for $\{Z\}$.

$$\frac{X \Rightarrow A}{\{X\} \Rightarrow \lceil A \rfloor} \lceil\,\rfloor \mathrm{R} \qquad \frac{X \Rightarrow A \qquad Y\{B\} \Rightarrow C}{Y\{B/A, X\} \Rightarrow C} \{/\}\mathrm{L} \qquad \frac{X \Rightarrow A \qquad Y\{B\} \Rightarrow C}{Y\{X, A\backslash B\} \Rightarrow C} \{\backslash\}\mathrm{L}$$

Fig. 7. Rules for brackets ⌈ ⌋ and structured { } sequents

$$\frac{[A], Z_1, [Z_2, B], W \Rightarrow C}{[A], Z_1, [Z_2, B|A], W \Rightarrow C} [|]\mathrm{L}_a \qquad \frac{[X_1, A, X_2], Z_1, [Z_2, B], W \Rightarrow C}{[X_1, A, X_2], Z_1, [Z_2, B\|A], W \Rightarrow C} [\|]\mathrm{L}_a$$

Fig. 8. Rules for (subject-oriented) reflexives and bound pronouns within a [ ] domain

$$\frac{[X_1, A, X_2], Z_1, \{Z_2, [Z_3, B], W\} \Rightarrow C}{[X_1, A, X_2], Z_1, \{Z_2, [Z_3, B\|A], W\} \Rightarrow C} \{|\}\,\mathrm{L}_a$$

$$\frac{[X_1, A, X_2], Z_1, \{Z_2, [Z_3, B], W\} \Rightarrow C}{[X_1, A, X_2], Z_1, \{Z_2, [Z_3, B\|A], W\} \Rightarrow C} \{\|\}\,\mathrm{L}_a$$

Fig. 9. Rules for bound pronouns and reflexives within a { } domain

We propose, consequently, bracketed versions for the $|\mathrm{L}_a$ and $\|\mathrm{L}_a$ rules, with the following side conditions: $X_1 \neq \epsilon$ or $X_2 \neq \epsilon$ in $[\|]\mathrm{L}_a$; $Z_3 \neq \epsilon$ in $\{|\}\mathrm{L}_a$.[15]

[15] As an anonymous reviewer pointed out, if the type $(\langle n \rangle \backslash s)/s$ were assigned to ECM verbs to differentiate them from propositional verbs, it would allow for ungrammatical sentences like (i–ii) below. To block binding of a reflexive in an object position by a non-local antecedent it seems we would have to impose some condition on the sequence Z_1 in $[\|]\mathrm{L}_a$. We plan to address the challenge posed by ECM constructions in future investigations.

(i) * John$_1$ expects Mary to like himself$_1$.
(ii) * John$_1$ believes Mary to expect Susan to like himself$_1$.

In addition, the side conditions on the $[\|]\mathrm{L}_a$ rule inadequately license pronouns to be bound by an antecedent within a conjunctive nominal phrase, as exemplified below. Indeed, *Mary* is taken as an argument of the functional type commonly assigned to *and*:

(iii) * John and Mary$_1$ praised her$_1$.
(iv) John$_1$ and Mary talked about him$_1$.

It appears that the unbracketed type assigned to the conjunction *and* has to be differentiated from the (bracketed) functional types assigned, for example, to *of*—$(n\backslash n)/\langle n \rangle$ and *'s*—$\langle n \rangle \backslash (n/cn)$. A distinction between a collective and a distributive type for *and* also seems to be relevant: roughly, $X\backslash X/X$ and $\langle X \rangle \backslash X/\langle X \rangle$, for example. For reasons of space, and since judgments seem to vary among speakers and sentences, we defer this problem to future research.

$$\dfrac{\dfrac{\dfrac{n \Rightarrow n}{[n] \Rightarrow \langle n\rangle}\langle\,\rangle\mathrm{R} \qquad \dfrac{\dfrac{n \Rightarrow n}{[n] \Rightarrow \langle n\rangle}\langle\,\rangle\mathrm{R} \quad \dfrac{s \Rightarrow s}{\{s\} \Rightarrow \lceil s\rfloor}\lceil\,\rfloor\mathrm{R}}{\{[n], \langle n\rangle \backslash s\} \Rightarrow \lceil s\rfloor}\{\backslash\}\mathrm{L}}{\{[n], ((\langle n\rangle \backslash s)/\langle n\rangle, [n]\} \Rightarrow \lceil s\rfloor}\{/\}\mathrm{L} \qquad \dfrac{\vdots}{[n], \langle n\rangle \backslash s \Rightarrow s}}{\dfrac{[n], ((\langle n\rangle \backslash s)/\lceil s\rfloor, \{[n], ((\langle n\rangle \backslash s)/\langle n\rangle, [n]\} \Rightarrow s}{[n], ((\langle n\rangle \backslash s)/\lceil s\rfloor, \{[n], ((\langle n\rangle \backslash s)/\langle n\rangle, [n\|n]\} \Rightarrow s}\{\|\}\mathrm{L}_a}/\mathrm{L}$$

Fig. 10. Derivation for *John*$_1$ *says Mary hates him*$_1$

$$\dfrac{\dfrac{\dfrac{\dfrac{\vdots}{[n], ((\langle n\rangle \backslash s)/\langle n\rangle, [n] \Rightarrow s}}{[n], ((\langle n\rangle \backslash s)/\langle n\rangle, [n|n] \Rightarrow s}[|]\mathrm{L}_a}{\{[n], ((\langle n\rangle \backslash s)/\langle n\rangle, [n|n]\} \Rightarrow \lceil s\rfloor}\lceil\,\rfloor\mathrm{R} \qquad \dfrac{\vdots}{[n], \langle n\rangle \backslash s \Rightarrow s}}{[n], \langle n\rangle \backslash \lceil s\rfloor, \{[n], ((\langle n\rangle \backslash s)/\langle n\rangle, [n|n]\} \Rightarrow s}/\mathrm{L}$$

Fig. 11. Derivation for *Mary says John hates himself*

$$\dfrac{\dfrac{\dfrac{\vdots}{\{[n], ((\langle n\rangle \backslash s)/\langle n\rangle, [n]\} \Rightarrow \lceil s\rfloor} \qquad \dfrac{\vdots}{[n], \langle n\rangle \backslash s \Rightarrow s}}{[n], ((\langle n\rangle \backslash s)/\lceil s\rfloor, \{[n], ((\langle n\rangle \backslash s)/\langle n\rangle, [n]\} \Rightarrow s}/\mathrm{L}}{[n], ((\langle n\rangle \backslash s)/\lceil s\rfloor, \{[n], ((\langle n\rangle \backslash s)/\langle n\rangle, [n|n]\} \Rightarrow s}\{|\}\mathrm{L}_a{*}$$

Fig. 12. Illicit derivation for *John*$_1$ *says Peter hates himself*$_1$

The rule $[|]\mathrm{L}_a$ in Fig. 8 preserves the prominence condition on the binder for reflexives: given that the reflexive within an argument domain $[\,]$ takes $[A]$ as its binder, the binder itself is not part of the subject (i.e. higher) argument. Conversely, the side conditions on the sequences X_1 and X_2 in $[\|]\mathrm{L}_a$ impede binding of a pronoun in an argument position if the binder is not part of the higher subject argument.

Note that there is no condition on the sequences X_1 and X_2 in the rule for pronouns $\{\|\}\mathrm{L}_a$ in Fig. 9. Thus, a pronoun within a propositional complement can be bound by a matrix subject (see Fig. 10). The side condition on the reflexive $\{|\}\mathrm{L}_a$ rule ensures that the reflexive stands in an embedded position within the propositional complement clause (contrast Figs. 11 and 12).

Nominal Complements. In nominal complements complementary distribution is fully verified: where the anaphora and its antecedent are co-arguments of the same function, a bound pronoun is ruled out. Nevertheless, a pronoun in the object complement position can still be bound provided that the binder itself is an argument of another functional type, as exemplified below:

(23) [John$_1$'s father]$_2$ loves him$_1$/himself$_2$.

(24) [The father of John$_1$]$_2$ loves him$_1$/himself$_2$.

From the following (bracketed) lexicon we may obtain the correct binding relations for reflexives and pronouns within the (direct) object argument position, as exemplified in (23–24) above and in (25–28) below (Fig. 13):

see/like : $(\langle n \rangle \backslash s)/\langle n \rangle$
john/mary : n
father/picture : cn
the/a : n/cn
of : $(n\backslash n)/n$
's : $n\backslash(n/cn)$

(25) John likes himself/*him.

(26) John takes a picture of himself/*him.

(27) * John saw himself's mother.

(28) John believes himself/*him to kiss Mary.

$$\frac{\begin{array}{c}\vdots\\ [n], (\langle n \rangle \backslash s)/\langle n \rangle, [n] \Rightarrow s\end{array}}{[n], (\langle n \rangle \backslash s)/\langle n \rangle, [n|n] \Rightarrow s}\ [|]\mathrm{L}_a/\ [||]\mathrm{L}_a*$$

Fig. 13. Derivation for $John_1$ $likes$ $himself_1/*him_1$

Prepositional Phrases. In general terms, scholars agree that prepositional phrases (PPs) selected by a verb can only contain a reflexive but not a bound pronoun, while prepositional phrases operating as adjuncts allow both a reflexive and a bound pronoun (cf. [4]).

Since our proposal strongly depends on the syntactic types assigned to the lexical items into the lexicon, the correctness of our proposal for anaphoric items within prepositional phrases mainly rests on the type assigned to the different classes of verbs.

Unfortunately, the distinction between complement prepositional phrases and adjunct phrases is not so pure in some cases. As claimed in [10], locative PPs, including those selected by a verb, must be distinguished from other PPs. Those verbs that select a PP bearing a locative role like *put* and *sit*, allow several locative prepositions, such as *in*, *on*, *near*, *into*, *next*, *in front of*. In this sense, locative PPs resemble adjunct PPs. By contrast verbs like *relies*, despite selecting a PP as complement, also select some specific preposition. The PP headed by *on/upon* in *relies on/upon* does not bear a locative role. Given this, it seems clear that we need to set a distinction between the PP selected by verbs like *put* and the PP selected by verbs like *relies*. In other terms, we need to set a bipartition into the set of PP complements: locative PPs and non-locative PPs. By using the bracket modality $\langle\ \rangle$ we mark the non-locative PP complement position into the lexical entry of the corresponding verbs and, taking into consideration the similarity between adjunct PPs and locative PP complements, we leave the PP position for the locative complement unmarked.[16]

Given that reflexives and anaphoric pronouns can occur within an unmarked position, we assume the following rules to process them (Fig. 14):

$$\frac{[X_1, A, X_2], Z, B, W \Rightarrow C}{[X_1, A, X_2], Z, B|A, W \Rightarrow C} |\mathrm{L}_a \qquad \frac{[X_1, A, X_2], Z, B, W \Rightarrow C}{[X_1, A, X_2], Z, B\|A, W \Rightarrow C} \|\mathrm{L}_a$$

Fig. 14. Rule for reflexives and anaphoric pronouns out of bracketed domains

Thus, assuming the following lexicon, we obtain the correct binding relation in different prepositional phrases, as exemplified in (29–32):

put/see : $((\langle n \rangle \backslash s)/pp)/\langle n \rangle$
glance : $(\langle n \rangle \backslash s)/pp$
rely : $(\langle n \rangle \backslash s)/\langle pp \rangle$
on/upon/behind/next : pp/n

(29) John_1 relies on himself_1/*him_1.

(30) John_1 glanced behind $\text{himself}_1/\text{him}_1$.

(31) John_1 put the gun near/under/on $\text{himself}_1/\text{him}_1$.

(32) John_1 saw a gun near $\text{himself}_1/\text{him}_1$.

4.2 Object-Oriented Reflexives and Bound Pronouns

Nominal Complements. Verbs like *show, give, send, promise, introduce* may select two nominal phrases as complements, and thus give rise to double-object constructions. These structures allow then for another pattern of reflexivization: reflexives bound by a nominal within a verb complement position. In other terms, besides subject-oriented reflexives, double-object constructions also allow for object-oriented ones. Double-object constructions alternate with oblique dative structures:

(33) Mary showed/gave/sent/promised John a gift.

(34) Mary showed/gave/sent/promised a gift to John.

[16] Alternatively, Reinhart and Reuland [22] consider that *relies on* forms a complex (semantic and syntactic) unit selecting a nominal complement, whilst *put* selects a prepositional complement. In view of this fact, we would assign the type $(\langle n \rangle \backslash s)/\langle n \rangle$ to *relies on/upon*.

Although the two structures display a different linear order, both reveal the same behavior when licensing anaphors. They show an asymmetry with respect to the licensing of object-oriented reflexives. As shown in the examples below, the correct binding relation for object-oriented reflexives in both structures depends on an ordering of the complements.

(35) Mary showed/presented John_1 himself_1.

(36) Mary showed/presented John_1 to himself_1.

Thus, these structures indicate that a hierarchical order between the two objects has to be imposed. To deal with double-object constructions we extend the calculus by adding a new product type-constructor.[17] We also present a different inference rule for object-oriented reflexives, where Z_1 does not contain a subtype s (Figs. 15 and 17).[18]

$$\frac{[X] \Rightarrow A \qquad [Y] \Rightarrow B}{[X, [Y]] \Rightarrow A \oslash B}\,[\oslash]\mathrm{R}$$

Fig. 15. Right rule for non-commutative $\oslash$ asymmetrical product

The correct binding relation is ensured by the following lexical assignment:
show/give/present/send : $(\langle n \rangle \backslash s)/(\langle n \rangle \oslash \langle n \rangle)$
show/give/present/send : $(\langle n \rangle \backslash s)/(\langle n \rangle \oslash \langle pp \rangle)$

[17] Note that the product $\oslash$ is not a discontinuous (or wrapping) type-constructor, unlike that of [1] or [19]. Since $\oslash$ is non-commutative, we would not be able to derive cases of "heavy" NP, as exemplified below. Nevertheless, in the following section we shall adopt a commutative product-type for the treatment of prepositional phrases.

(i) I gave to the students presents that I had brought back from Spain.

To deal with double-object structures, Hepple [6] extends the **L** calculus by adding a new slash type-constructor $\oint$ and a modality $\triangleright$. Since the slash type-constructor lacks introduction rules, it may encode the hierarchical ordering of the nominal complements; the modality allows the nominal complements to be reordered to obtain the correct surface word-order.

[18] We note that a slightly modified version of the rule in Fig. 16 may also be used for anaphors in a complement of possessives, which are not either subject nor object-oriented. Once again, it appears that a distinction between the functional type assigned to *of* or *'s* and *and* has to be made to prevent *He and himself* from assigning the type n.

(i) John_1's description of himself_1 is lovely.

(ii) Lisa burned Andy Warhol_1's portrait of himself_1.

It seems that it could be possible to also encode a hierarchical ordering into the rules for the Lambek slash type-constructors. In this case, it would be possible to deal with subject- and object-oriented anaphors in a uniform way.

$$\frac{X_1, [X_2, A, Z_1, [Z_2, B]], W \Rightarrow C}{X_1, [X_2, A, Z_1, [Z_2, B|A]], W \Rightarrow C}\, [|]\mathrm{L}_a$$

Fig. 16. Rule for object-oriented reflexives

$$\cfrac{\cfrac{\cfrac{[n] \Rightarrow \langle n\rangle \qquad [n] \Rightarrow \langle n\rangle}{[n, [n]] \Rightarrow \langle n\rangle \otimes \langle n\rangle}\,\otimes\mathrm{R} \qquad \cfrac{\vdots}{[n], \langle n\rangle \backslash s \Rightarrow s}}{[n], (\langle n\rangle \backslash s)/(\langle n\rangle \otimes \langle n\rangle), [n, [n]] \Rightarrow s}\,/\mathrm{L}}{[n], (\langle n\rangle \backslash s)/(\langle n\rangle \otimes \langle n\rangle), [n, [n|n]] \Rightarrow s}\,[|]\mathrm{L}_a$$

Fig. 17. Derivation for *Mary presents John himself*

Prepositional Phrases: The *About*-Phrase. Verbs selecting two prepositional phrases also challenge several binding theories. In this case, there is also no complete agreement among scholars with respect to their syntactic status.[19] As is known, two prepositional complements may appear in either order:

(37) John talked to Mary about Bill.

(38) John talked about Bill to Mary.

Despite the free word-order, the occurrence of a reflexive within a prepositional phrase, such as in double-object structures, indicates that a structural ordering between the *about*-phrase and the *to*-phrase has to be imposed.

(39) John talked to Mary$_1$ about herself$_1$.

(40) * John talked about Mary$_1$ to herself$_1$.

[19] In some generative theories, the *about*-phrase is evaluated as an adjunct phrase and thus is separated from the *to*-phrase or *with*-phrase complement (cf. [22]). This would explain the ungrammaticality of (40), but not the ungrammaticality of (i) below.

(i) * Mary talked to John$_1$ about him$_1$.

In other theories, the *about*-phrase, as well as the *to*-phrase, is considered a verb complement; the difference between these PPs is made by assuming an ordering with respect to their relative obliqueness: the *about*-phrase is more oblique than the *to*-phrase (cf. [20]). Since the anaphor has to be bound by a less oblique co-argument, the relationship of relative obliqueness would account for (i) above, but not for (ii) below, where the linear word-order seems to be also relevant.

(ii) * Mary talked about himself$_1$ to John$_1$.

In addition, [3] suggests an approach in which the verb *talk* (and also *speak*) and the preposition *to* are reanalyzed as one verb taking a nominal object (and a prepositional complement) (cf. also [23]). Thus, *talk* would be analogous to (one of the forms of) *tell*. To formalize this proposal, besides encoding free linear word-order and relative obliqueness, the syntactic functional type assigned to the *talk to*-phrase would have to encode discontinuity as well.

$$\frac{[X] \Rightarrow A \qquad [Y] \Rightarrow B}{[X,[Y]] \Rightarrow A \otimes B}\,[\otimes]R_1 \qquad \frac{[X] \Rightarrow B \qquad [Y] \Rightarrow A}{[[X],Y] \Rightarrow A \otimes B}\,[\otimes]R_2$$

Fig. 18. Right rules for commutative asymmetrical product $\otimes$

$$\dfrac{\dfrac{\dfrac{[PP_{about}/n, n] \Rightarrow \langle PP_{about}\rangle \qquad [PP_{to}/n, n] \Rightarrow \langle PP_{to}\rangle}{[[PP_{about}/n, n], PP_{to}/n, n] \Rightarrow \langle PP_{to}\rangle \otimes \langle PP_{about}\rangle}\otimes R_2 \qquad [n], \langle n\rangle \backslash s \Rightarrow s}{[n], (\langle n\rangle \backslash s)/(\langle PP_{to}\rangle \otimes \langle PP_{about}\rangle), [[PP_{about}/n, n], PP_{to}/n, n] \Rightarrow s}/L}{[n], (\langle n\rangle \backslash s)/(\langle PP_{to}\rangle \otimes \langle PP_{about}\rangle), [[PP_{about}/n, n], PP_{to}/n, n|n] \Rightarrow s}[|]L_a*$$

Fig. 19. Illicit derivation for **John talked about Mary to herself*

$$\dfrac{\dfrac{\dfrac{[PP_{to}/n, n] \Rightarrow \langle PP_{to}\rangle \qquad [pp_{about}/n, n] \Rightarrow \langle pp_{about}\rangle}{[pp_{to}/n, n, [pp_{about}/n, n]] \Rightarrow \langle pp_{to}\rangle \otimes \langle pp_{about}\rangle}\otimes R_1 \qquad [n], \langle n\rangle \backslash s \Rightarrow s}{[n], (\langle n\rangle \backslash s)/(\langle pp_{to}\rangle \otimes \langle pp_{about}\rangle), [pp_{to}/n, n, [pp_{about}/n, n]] \Rightarrow s}/L}{[n], (\langle n\rangle \backslash s)/(\langle pp_{to}\rangle \otimes \langle pp_{about}\rangle), [pp_{to}/n, n, [pp_{about}/n, n|n]] \Rightarrow s}[|]L_a$$

Fig. 20. Derivation for *John talked to Mary about herself*

In a categorial framework, it is the functional type assigned to a verb like *talk* which has to express the different syntactic relation that these two PPs maintain with the verb. In [16], for example, the type assigned to *talk* is $((n\backslash s)/pp)/pp$, while in [19] it is the type $(n\backslash s)/(pp_{to} \otimes pp_{about})$, where $\otimes$ is the nondeterministic continuous product of the Displacement Calculus **D**. Thus, this last type captures the alternative surface word-order. This notwithstanding, by using either the former or the latter type, prepositional phrases both get the same syntactic non-hierarchical status of verb complements.[20]

Hence, the different hierarchical relation the PPs complements maintain with the verb seems to call for a new type-constructor that is analogous to that we have used to deal with double-object constructions, but which encodes commutativity as well (Figs. 18, 19 and 20).

With this type-constructor at hand, we then propose the following lexical assignment:

talk : $(\langle n\rangle \backslash s)/(\langle pp_{to}\rangle \otimes \langle pp_{about}\rangle)$

4.3 Long-Distance Anaphors

Anaphors in Icelandic are necessarily subject-oriented and do not respect Principle A for anaphors, as they can be bound by a long-distance antecedent, provided

[20] Since the calculus **D** also contains a nondeterministic discontinuous product $\odot$, the type $(n\backslash s)/(pp \odot pp)$ would take the structural ordering into account if the premisses of the right rule were bracketed sequences.

that the anaphora stands in a subjunctive clause. In this sense, long-distance anaphors resemble anaphoric pronouns in propositional (finite) complements from English. In addition, the subjunctive mood in Icelandic may be propagated down through embedded complements (this is the so-called *domino effect*). Given that bracket modalities have been applied in Type-Logical Grammar to delimit syntactic domains, we suggest using the bracket $\{\ \}$ to simulate the domino effect of the subjunctive mood generated by some verbs (e.g. *segir* 'say' vs. *víta* 'know') and the bracket $[\ ]$ to ensure binding only by the subject (that is, the subject condition; cf. [18]). Since the licensing of a long-distance anaphor in this language also depends on the case properties of the binder and bindee, we merely sketch an analysis here. The left rule for long distance anaphors is given in Fig. 21.

segir 'say': $((\langle n\rangle\backslash s)/\lceil s\rfloor$
víta 'know': $((\langle n\rangle\backslash s)/s$
elskar 'love': $((\langle n\rangle\backslash s)/n$

(41) Jón$_1$ segir að María$_2$ elski sig$_1$.
John say that Maria love.SUBJ SE-ANAPHOR.ACC
'John says that Mary loves him.'

(42) ?Jón$_1$ veit að María elskar sig$_1$.
John know that Maria love.IND SE-ANAPHOR.ACC
'John knows that Mary loves him.'

$$\frac{X, [A], Z_1, \{Z_2, B, W\} \Rightarrow C}{X, [A], Z_1, \{Z_2, B|A, W\} \Rightarrow C}\ |\mathrm{L}_{lg}$$

Fig. 21. Rule for long-distance anaphor

5 Conclusions

In this paper we have proposed different rules to deal with anaphoric and pronominal pronouns occurring in several syntactic domains. Although both the type assignment for pronouns and our initial idea for the pronominal rules come from Jäeger [8], we have proposed a different type assignment for reflexives and pronominal pronouns and we have modified the rules of **LLC**. The inspiration for lexical entries encoding marked argument positions comes from [18]. By adopting bracket modalities we have identified different syntactic domains; in light of the latter, we have encoded binding restrictions into the left anaphoric rule of **LLBE**. The right pronominal rule of **LLBE**, in turn, evidences that despite the fact that an antecedent A could occur in the local syntactic domain $[\ ]$, a free pronoun is derived by assuming an antecedent in a non-local domain $[\![\ ]\!]$.

The rules of **LLBE** reveal then that free and anaphoric pronouns on the one hand, and bound pronouns on the other, are generally processed in different steps in a proof: if $X_1 = X_2 = \epsilon$ and so the antecedent A is left-peripheral, free pronouns and reflexives, but not bound pronouns, can be inserted into a derivation. Our proposal preserves the prominence condition on the binder for reflexives: the binder may not be an argument lower in the hierarchy and neither may it be part of an argument higher in the hierarchy. In addition, we have incorporated the previous modal categorial analysis for *say* in terms of structural modalities, in accordance with our overall proposal. Further, we have suggested how this proposal can be used to deal with long-distance anaphors in Icelandic. Our rule for object-oriented reflexives could also be used to deal with non-subject- and non-object-oriented reflexives, such as anaphors in possessive complements.

In future work, we propose to investigate how to impose structural conditions upon the sequences of the left rules for the customized slash type-constructors in order to reduce the number of pronominal rules and thus to deal with subject and object anaphoric pronouns in a more uniform way. We also plan to explore how to deal with ECM constructions.

Acknowledgment. The author was supported by a doctoral scholarship granted by FAPESP (Fundação de Amparo à Pesquisa do Estado de São Paulo, process number 2013/08115-1).

Appendix

The proof for the Cut elimination theorem requires the use of the following bracketed versions of the structural rules of Permutation and Expansion (Fig. 22). In order to prove Cut Elimination for **LLBE** we have to consider two more cases for principal Cut: the left premise of Cut is the conclusion of $\|\mathrm{L}_a$ or that of $\|\mathrm{L}_p$ and the right premise is the conclusion of $\|\mathrm{R}$. These two configurations are given schematically in Figs. 23 and 24. In both cases, the principal Cut is replaced by a Cut of lower degree. Since no rule introduces a formula $[\![A]\!]$ into the right side of a sequent (i.e. there are only antecedent occurrences of the formula $[\![A]\!]$), the Cut formula could not have been derived by applying either of the bracketed structural rules.

$$\frac{X, A, Y, Z \Rightarrow C}{X, A, Y, [\![A]\!], Z \Rightarrow C}\,[\![E]\!] \qquad \frac{X, [\![A]\!], Y, Z \Rightarrow C}{X, Y, [\![A]\!], Z \Rightarrow C}\,[\![P]\!]$$

Fig. 22. Bracketed structural rules

$$\frac{\dfrac{Z_1 \Rightarrow C}{\llbracket A \rrbracket, Z_1 \Rightarrow C \| A} \|R \qquad \dfrac{X_2, A, Z_2, C, W_2 \Rightarrow D}{X_2, A, Z_2, C \| A, W_2 \Rightarrow D} \|\mathrm{L}_a}{X_2, A, Z_2, \llbracket A \rrbracket, Z_1, W_2 \Rightarrow D} Cut$$

$$\rightsquigarrow$$

$$\frac{\dfrac{Z_1 \Rightarrow C \qquad X_2, A, Z_2, C, W_2 \Rightarrow D}{X_2, A, Z_2, Z_1, W_2 \Rightarrow D} Cut}{X_2, A, Z_2, \llbracket A \rrbracket, Z_1, W_2 \Rightarrow D} \llbracket E \rrbracket$$

Fig. 23. Principal cut for $\|$: $\|\mathrm{L}_a$

$$\frac{\dfrac{Z_1 \Rightarrow C}{\llbracket A \rrbracket, Z_1 \Rightarrow C \| A} \|\mathrm{R} \qquad \dfrac{X_2, \llbracket A \rrbracket, Z_2, C, W_2 \Rightarrow D}{X_2, Z_2, C \| A, W_2 \Rightarrow D} \|\mathrm{L}_p}{X_2, Z_2, \llbracket A \rrbracket, Z_1, W_2 \Rightarrow D} Cut$$

$$\rightsquigarrow$$

$$\frac{\dfrac{Z_1 \Rightarrow C \qquad X_2, \llbracket A \rrbracket, Z_2, C, W_2 \Rightarrow D}{X_2, \llbracket A \rrbracket, Z_2, Z_1, W_2 \Rightarrow D} Cut}{X_2, Z_2, \llbracket A \rrbracket, Z_1, W_2 \Rightarrow D} \llbracket P \rrbracket$$

Fig. 24. Principal cut for $\|$: $\|\mathrm{L}_p$

References

1. Bach, E.: Control in montague grammar. Linguist. Inq. **10**(4), 515–531 (1979)
2. Büring, D.: Pronouns. In: Semantics: An International Handbook of Natural Language Meaning, vol. 2, pp. 971–996 (2011)
3. Chomsky, N.: Lectures on Government and Binding. Kluwer, Dordrecht (1981)
4. Everaert, M.: Binding theories: a comparison of grammatical models. In: van Oostendorp, M., Anagnostopoulou, E. (eds.) Progress in Grammar. Articles at the 20th Anniversary of the Comparison of Grammatical Models Group in Tilburg. Meertens Institute, Electronic Publications in Linguistics, Amsterdam (2000)
5. Hendriks, P., Hoeks, J., Spenader, J.: Reflexive choice in Dutch and German. J. Comp. Ger. Linguist. **17**(3), 229–252 (2015)
6. Hepple, M.: Command and domain constraints in a categorial theory of binding. In: Proceedings of the Eight Amsterdam Colloquium, pp. 253–270 (1992)
7. Jacobson, P.: Towards a variable-free semantics. Linguist. Philos. **22**(2), 117–185 (1999)
8. Jäeger, G.: Anaphora and Type Logical Grammar. Trends in Logic - Studia Logica Library, vol. 24. Springer, Dordrecht (2005). https://doi.org/10.1007/1-4020-3905-0
9. Lambek, J.: The mathematics of sentence structure. Am. Math. Mon. **65**(3), 154–170 (1958)
10. Marantz, A.P.: On the Nature of Grammatical Relations. Linguistic Inquiry Monographs Ten. The MIT Press, Cambridge (1984)
11. Mensching, G.: Infinitive constructions with specified subjects: a syntactic analysis of the romance languages (2000)

12. Moortgat, M.: Multimodal linguistic inference. J. Logic Lang. Inform. **5**(3–4), 349–385 (1996)
13. Morrill, G.: Intensionality and boundedness. Linguist. Philos. **13**(6), 699–726 (1990)
14. Morrill, G.: Categorial formalisation of relativisation: pied piping, islands, and extraction sites. Technical report, Departament de Llenguatges i Sistemes Informàtics, Universitat Politècnica de Catalunya (1992)
15. Morrill, G.: Type Logical Grammar. Categorial Logic of Signs. Springer, Dordrecht (1994). https://doi.org/10.1007/978-94-011-1042-6
16. Morrill, G., Valentín, O.: On anaphora and the binding principles in categorial grammar. In: Dawar, A., de Queiroz, R. (eds.) WoLLIC 2010. LNCS (LNAI), vol. 6188, pp. 176–190. Springer, Heidelberg (2010). https://doi.org/10.1007/978-3-642-13824-9_5
17. Morrill, G., Valentín, O.: Semantically inactive multiplicatives and words as types. In: Asher, N., Soloviev, S. (eds.) LACL 2014. LNCS, vol. 8535, pp. 149–162. Springer, Heidelberg (2014). https://doi.org/10.1007/978-3-662-43742-1_2
18. Morrill, G., Valentín, O.: Computational coverage of TLG: displacement. In: Proceedings of Empirical Advances in Categorial Grammar, pp. 132–161 (2015)
19. Morrill, G., Valentín, O., Fadda, M.: The displacement calculus. J. Logic Lang. Inform. **20**(1), 1–48 (2011)
20. Pollard, C., Sag, I.A.: Anaphors in English and the scope of binding theory. Linguist. Inq. **23**(2), 261–303 (1992)
21. Reinhart, T.: Coreference and bound anaphora: a restatement of the anaphora questions. Linguist. Philos. **6**(1), 47–88 (1983)
22. Reinhart, T., Reuland, E.: Reflexivity. Linguist. Inq. **24**(4), 657–720 (1993)
23. Szabolcsi, A.: Bound variables in syntax (are there any?). In: Bartsch, R., van Benthem, J., van Emde Boas, P. (eds.) Semantics and Contextual Expressions, pp. 295–318. Foris, Dordrecht (1989)

Morphological Agreement in Minimalist Grammars

Marina Ermolaeva(✉)

University of Chicago, Chicago, USA
mermolaeva@uchicago.edu

Abstract. Minimalist Grammars provide a useful tool for modeling natural language syntax by defining grammar fragments in a very precise way. As a formalization of the Minimalist Program, they can accommodate linguistic analyses from the field of generative syntax. However, they have no machinery for encoding agreement; while morphology can be simulated by multiplying lexical items, there is no systematic way to state generalizations and implement actual proposals. This paper extends Minimalist Grammars with morphological features and operations on them. As a proof of concept, I show how Icelandic dative intervention can be encoded in the modified formalism.

Keywords: Minimalist Grammars · Minimalist Syntax · Agreement Morphosyntax · Icelandic

1 Introduction

Agreement can be defined as the morphological manifestation of dependencies between words. In a basic English sentence like *He walks* the verb agrees in person and number with the subject, and the subject, in turn, receives nominative case from the verb.[1] These dependencies may be nonlocal; for instance, English expletive constructions like *There seems to have arrived a man* exhibit long-distance subject-verb agreement.

Chomsky's Minimalist Program [4,5] treats these phenomena as an effect of a much more general mechanism known as Agree. An explicit theory of feature structures compatible with Chomsky's framework is proposed by Adger in [1]. Lexical items are defined as sets of features, each specified as bearing a value (drawn from some finite set) or being unvalued. Syntax is driven by features: the *probe* of a syntactic operation is an element with an unvalued feature, and the *goal* must bear a matching valued feature. Adger defines feature *valuation* as unification of values (cf. [15]): the unvalued feature on the probe assumes the value of the goal. The three operations are Merge, Move, and Agree. Merge and Move operate on *categorial features* (T, V, N ...) and build new structure. Agree targets *morphosyntactic features* (case, number, person, ...) and forms dependencies between elements of the existing structure.

[1] For the sake of exposition, I assume that case assignment reduces to agreement and that structural case is explicitly assigned by finite verbs. Neither is free of controversy.

A. Foret et al. (Eds.): FG 2017, LNCS 10686, pp. 20–36, 2018.
https://doi.org/10.1007/978-3-662-56343-4_2

Example 1. The phrase *this nice boy* exhibits determiner-noun agreement. The determiner bears an unvalued number feature and dominates the noun, which has a valued feature. This probe-goal configuration allows Agree to apply:

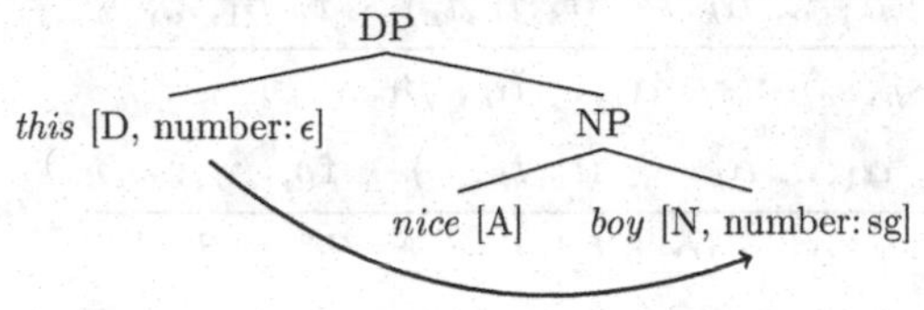

Stabler's Minimalist Grammars (MGs, [16,17]) have been designed as a mathematically rigorous formalization of Minimalist Syntax. The MG formalism is based on operations analogous to Merge and Move. Agree, however, has no counterpart. My goal is to extend MGs in a way that retains the relation to the Minimalist Program, allowing to translate Minimalist proposals involving agreement into the modified formalism.

2 Minimalist Grammars

I begin with the version of Minimalist Grammars defined in [17], with a few tweaks. This formalism employs *chain notation*, reducing syntactically active subtrees of derived trees to tuples of strings.

Definition 2. A minimalist grammar G is a 5-tuple $\langle \Sigma, Syn, Types, Lex, \mathcal{F} \rangle$, where

Σ is a finite set (of pronounced segments),

$Syn = Base$ (nonempty finite set of *categories*)
$\quad \cup \{\texttt{=f} \mid \texttt{f} \in Base\} \cup \{\texttt{=>f} \mid \texttt{f} \in Base\}$ (*selectors*)
$\quad \cup \{\texttt{+f} \mid \texttt{f} \in Base\}$ (*licensors*)
$\quad \cup \{\texttt{-f} \mid \texttt{f} \in Base\} \cup \{\texttt{*f} \mid \texttt{f} \in Base\}$ (*licensees*)
is a set of syntactic features,

$Types = \{::, :\}$, (lexical, derived)

Let the set of *initial chains*[2] $IC = \Sigma^* \times \Sigma^* \times \Sigma^*\ Types\ Syn^*$, and the set of *non-initial chains* $NC = \Sigma^*\ Syn^*$;
$Lex \subset \{\epsilon\} \times \Sigma^* \times \{\epsilon\}\ \{::\}\ Syn^*$, a subset of IC, is a finite set of lexical items (*lexicon*),
$\mathcal{F} = \{merge, move\}$ is a set of structure-building operations:
- *merge* is the union of the following five functions, for $s_s, s_h, s_c, t_s, t_h, t_c \in \Sigma^*$, $\cdot \in \{:, ::\}$, $\texttt{f} \in Base$, $\gamma \in Syn^*$, $\delta \in Syn^+$, $\alpha_1, ..., \alpha_k, \beta_1, ..., \beta_l \in NC$ $(0 \leq k, l)$,

[2] Angle brackets are used to denote tuples. For any n-tuple or sequence, for $1 \leq i \leq n$, $T[i]$ denotes the ith component of T. The (finite) product of sets $A_1, A_2, ..., A_n$ $A_1 \times A_2 \times ... \times A_n = \{\langle a_1, a_2, ..., a_n \rangle \mid a_1 \in A_1, a_2 \in A_2, ..., a_n \in A_n\}$. Similarly, their concatenation $A_1 A_2 ... A_n = \{a_1 a_2 ... a_n \mid a_1 \in A_1, a_2 \in A_2, ..., a_n \in A_n\}$.

$$mrg1: \frac{\langle \epsilon, s_h, \epsilon \rangle :: \texttt{=f}\gamma \qquad \langle t_s, t_h, t_c \rangle \cdot \texttt{f},\ \beta_1, ..., \beta_l}{\langle \epsilon, s_h, t_s t_h t_c \rangle : \gamma,\ \beta_1, ..., \beta_l}$$

$$mrg2: \frac{\langle s_s, s_h, s_c \rangle : \texttt{=f}\gamma,\ \alpha_1, ..., \alpha_k \qquad \langle t_s, t_h, t_c \rangle \cdot \texttt{f},\ \beta_1, ..., \beta_l}{\langle t_s t_h t_c s_s, s_h, s_c \rangle : \gamma,\ \alpha_1, ..., \alpha_k,\ \beta_1, ..., \beta_l}$$

$$mrg3: \frac{\langle s_s, s_h, s_c \rangle \cdot \texttt{=f}\gamma,\ \alpha_1, ..., \alpha_k \qquad \langle t_s, t_h, t_c \rangle \cdot \texttt{f}\delta,\ \beta_1, ..., \beta_l}{\langle s_s, s_h, s_c \rangle : \gamma,\ \alpha_1, ..., \alpha_k,\ t_s t_h t_c : \delta,\ \beta_1, ..., \beta_l}$$

$$hmrg1: \frac{\langle \epsilon, s_h, \epsilon \rangle :: \texttt{=>f}\gamma \qquad \langle t_s, t_h, t_c \rangle \cdot \texttt{f},\ \beta_1, ..., \beta_l}{\langle \epsilon, t_h s_h, t_s t_c \rangle : \gamma,\ \beta_1, ..., \beta_l}$$

$$hmrg3: \frac{\langle s_s, s_h, s_c \rangle \cdot \texttt{=>f}\gamma,\ \alpha_1, ..., \alpha_k \qquad \langle t_s, t_h, t_c \rangle \cdot \texttt{f}\delta,\ \beta_1, ..., \beta_l}{\langle s_s, t_h s_h, s_c \rangle : \gamma, \alpha_1, ..., \alpha_k,\ t_s t_c : \delta,\ \beta_1, ..., \beta_l}$$

- *move* is the union of the following three functions, for $s_s, s_h, s_c, t \in \Sigma^*$, $\texttt{f} \in Base$, $\texttt{F} \in \{\texttt{-f}, \texttt{*f}\}$, $\gamma, \zeta \in Syn^*$, $\delta \in Syn^+$, and for $\alpha_1, ..., \alpha_k \in NC$ $(0 \leq k)$ satisfying the condition (SMC)[3] there is exactly one $i \in [1, k]$ such that α_i has `-f` or `*f` as its first feature,

$$mv1: \frac{\langle s_s, s_h, s_c \rangle \cdot \texttt{+f}\gamma,\ \alpha_1, ..., \alpha_{i-1},\ t\ \texttt{F},\ \alpha_{i+1}, ..., \alpha_k}{\langle t s_s, s_h, s_c \rangle : \gamma,\ \alpha_1, ..., \alpha_{i-1},\ \alpha_{i+1}, ..., \alpha_k}$$

$$mv2: \frac{\langle s_s, s_h, s_c \rangle \cdot \texttt{+f}\gamma,\ \alpha_1, ..., \alpha_{i-1},\ t\ \texttt{F}\delta,\ \alpha_{i+1}, ..., \alpha_k}{\langle s_s, s_h, s_c \rangle : \gamma,\ \alpha_1, ..., \alpha_{i-1},\ t : \delta,\ \alpha_{i+1}, ..., \alpha_k}$$

$$mv*: \frac{\langle s_s, s_h, s_c \rangle \cdot \texttt{+f}\gamma,\ \alpha_1, ..., \alpha_{i-1},\ t\ \texttt{*f}\zeta,\ \alpha_{i+1}, ..., \alpha_k}{\langle s_s, s_h, s_c \rangle : \gamma,\ \alpha_1, ..., \alpha_{i-1},\ t : \texttt{*f}\zeta,\ \alpha_{i+1}, ..., \alpha_k}$$

Definition 3. An *expression* is a member of $Exp = IC\ NC^*$. An expression e is a *complete expression* of category $\texttt{c} \in Base$ iff $e = \langle s_s, s_h, s_t \rangle \cdot \texttt{c}$, where $\cdot \in \{::, :\}$.

Starred licensees of the form $*$`f` are optionally deleted (by *mv1* or *mv2*) or saved for later (by *mv*∗). The latter possibility corresponds to intermediate positions of movement. [16] mentions this option of implementing successive cyclic movement; and a version of MGs with starred categorial features is explored in [11]. A formalism with persistent features, optionally erased by syntactic operations, has been shown to be weakly equivalent to standard MGs [18].

MGs offer a limited means of encoding (agglutinative) morphology by assigning separate lexical items to morphemes and constructing morphological words with head movement. Dependencies between words can be enforced by building restrictions into syntactic features.

[3] The SMC (Shortest Move Constraint): is a special case of the requirement that at any given step in the derivation the derived structure contain only finitely many subtrees (chains) which are syntactically active (i.e. have unchecked features).

Example 4. $G = \langle \Sigma_G, Syn_G, Types, Lex_G, \mathcal{F} \rangle$ is an MG. Its lexicon Lex_G contains the following lexical items:

$$\begin{array}{lll} \texttt{this.sg.nom} & := \langle \epsilon,\ \textit{this},\ \epsilon \rangle & :: \texttt{=n}_{\texttt{3SgN}}\ \texttt{d -k}_{\texttt{3SgN}} \\ \texttt{boy.sg.nom} & := \langle \epsilon,\ \textit{boy},\ \epsilon \rangle & :: \texttt{n}_{\texttt{3SgN}} \\ \texttt{walk} & := \langle \epsilon,\ \textit{walk},\ \epsilon \rangle & :: \texttt{=d v} \\ \texttt{prs.3sg} & := \langle \epsilon,\ \textit{-s},\ \epsilon \rangle & :: \texttt{=>v +k}_{\texttt{3SgN}}\ \texttt{t} \end{array}$$

G generates one expression of category t, derived as follows:

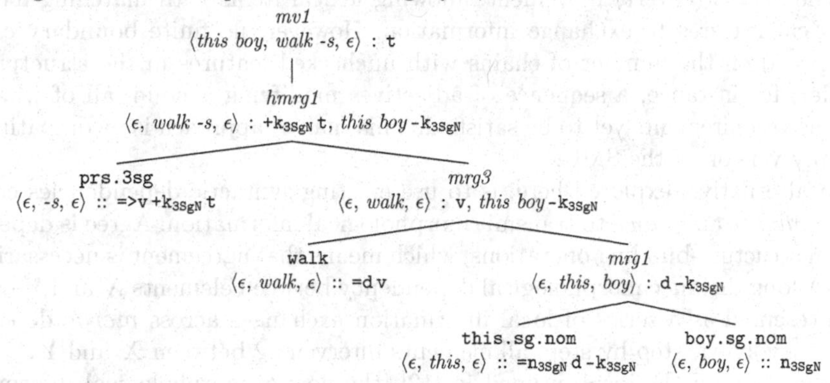

This toy grammar forces *merge* and *move* to only combine lexical items with compatible morphological features – at the expense of having a separate feature for each combination of morphological properties that may result in a distinct morphological form. All agreement in *this boy walk -s* is local, which makes it easy to refine syntactic features manually. For long-distance agreement dependencies, a better option is to state compatibility restrictions as constraints defined in monadic second-order logic, as shown in [8].

The generative capacity of MGs is sufficient to encode any mildly context-sensitive pattern, so this strategy is adequate for ensuring correct agreement. However, it does not provide a succinct, systematic way of formulating generalizations about morphological dependencies; the relation to the Minimalist Agree operation remains obscure. Furthermore, the mappings from derivation to pronounced form in MGs have been given without special attention to the complications imposed by a detailed model of morphology. In the next section I propose a refinement which allows for straightforward integration with standard models of morphology.

3 Towards Agreement

3.1 Bird's-Eye View

Bundles and Channels. MGs treat all features as uninterpretable – in the sense that they all (with the exception of one category feature) must be deleted to form a complete expression. Morphological agreement essentially requires a class of features which are valued in the course of derivation and serve as building blocks of syntactic output. The first step is to redefine lexical items, replacing each sequence of phonological segments with a *bundle* – a set of morphological features. Incidentally, this modification separates syntax from phonology: pronounced segments are no longer present in lexical items and are assumed to be inserted outside syntax.

What about feature valuation? One option is an almost faithful translation of Minimalist Agree [1] into the MG formalism. Agree can be straightforwardly implemented as (covert) movement, allowing lexical items with matching morphological features to exchange information. However, no finite boundary can be imposed on the number of chains with unchecked features in the structure: consider, for instance, a sequence of adjectives modifying a noun, all of which have case requirements yet to be satisfied. This "naive" approach is incompatible with any version of the SMC.

An alternative, explored here, is to use existing syntactic dependencies created by *merge* and *move* to transmit morphological information. Agree is dependent on structure-building operations, which means that agreement is necessarily local. A long-distance morphological dependency between elements X and Y can be represented as a series of local information exchanges across *merge* dependencies involving, step by step, all elements intervening between X and Y.

Expanding on the idea outlined in [12], the flow of morphological information can be controlled by annotating syntactic features with their agreement properties, which can be conveniently thought of in terms of *channels*. For each syntactic feature, one needs to specify whether it accepts information from whatever checks it (*receiving channel*) and which values it transmits to whatever checks it (*emitting channel*).[4] Whenever two syntactic features establish a syntactic dependency, and one of them has a receiving channel, the chain/expression bearing this feature is updated with values specified for the emitting channel of the other feature. Borrowing terminology from linguistic literature, I call this process *downward agreement* if the feature at the receiving end is a selector or licensor, and *upward agreement* if it is a categorial feature or licensee.

Example 5. Recall the lexical item `boy.sg` from Example 4. The phonological exponent *boy* is replaced with a bundle, with ϵ being the *default value*:

$$\texttt{boy.sg} := \left\langle \epsilon, \begin{bmatrix} \text{BOY} \\ \text{num:}\mathit{sg} \\ \text{per:}\mathit{3} \\ \text{case:}\epsilon \end{bmatrix}, \epsilon \right\rangle :: \texttt{n}_{\leftarrow}^{\begin{bmatrix} \text{num:}\mathit{sg} \\ \text{per:}\mathit{3} \end{bmatrix}\rightarrow}$$

The category feature `n` has a receiving channel (denoted by $_\leftarrow$), which allows `boy.sg` to receive a case value via upward agreement. The emitting channel on `n` (indicated by $^\rightarrow$) transmits number and person to whatever selects `boy.sg`.

[4] A lexical item may transmit different values of the same morphological feature via different channels. Moreover, these values need not be a subset of values in the item's own bundle. Keeping the content of emitting channels unconstrained is useful: for example, a preposition is allowed to transmit lexical case to its complement without morphologically manifesting it itself.

Probes and Goals. The channel-based agreement system can be refined to bring it more in line with the traditional notion of Agree. One such restriction is mentioned in [5] as a locality condition on goal defined in terms of "closest c-command" and in [1] as the requirement that the features in a probe-goal relation have no other matching feature intervening between them. The channel system has a built-in locality condition: each lexical item interacts directly with the head of the expression it selects/licenses and is not allowed to probe further. For items with multiple selectors/licensors, it is sufficient to require heads to accept agreement information via their *last* receiving channel. In other words, later values overwrite those received earlier. The intuition is simple: if X is selected by Y, the argument of X which is merged last will be the closest goal for Y.

Another useful restriction is known as the *freezing effect* of feature checking [2]. In essence, it rules out agreement in intermediate positions of successive cyclic movement. This condition is only relevant for starred licensees and can be built into the definition of *mv∗* in a straightforward way.

Feature Sharing. Long-distance upward agreement (across *merge* dependencies) cannot be reconciled with the requirement that the goal always provide a valued feature to the probe. When this is not the case, the agreement relation between lexical items in an expression has to be recorded so that, when the needed value enters the derivation, both items could be updated simultaneously. Nothing prevents such a relation from spanning multiple chains. The MG formalism distinguishes between the initial chain and non-initial chains but does not record any hierarchical relations. Additional bookkeeping is required to keep track of this information.

The proposed solution is reminiscent of *feature sharing* [7]. Their version of Agree does not require the goal feature to bear a value: matching features become a shared feature which is valued if either of the coalescing features is valued. I adopt a similar approach by recording for each feature, alongside its value, its *rewritability* – the highest chain that can transmit a value to it. With the SMC in place, every non-initial chain is uniquely identified by the name of its first licensee. Thus, rewritability can be set to `on` (*active*, or sharing value with the initial chain), a licensee name (sharing value with a non-initial chain), or `off` (*inactive*, or not accessible to agreement). I assume that morphological features start out as `off` if valued in the lexicon.

For any chain in an expression, its *subchains* are chains representing its subtrees. Each non-initial chain can be annotated with the sequence of all non-initial chains it is a subchain of (including itself), from the outermost to the most embedded. I will refer to this sequence as *lineage* of the chain. By convention, all lineages end with an `off` value. The set of all lineages for a given grammar is the set of all sequences of elements of *Base* without repetitions, followed by `off`. Lineages are updated throughout the derivation. Whenever a new non-initial chain appears in the expression, the name of its first licensee is prepended to the lineage of the new chain as well as all its subchains. On the other hand, when a chain moves, it ceases to be a subchain of any non-initial chain.[5] Therefore, all chains undergoing movement are stripped of the initial part of their lineage up to and including the relevant licensee.

Long-distance upward agreement succeeds if there is an uninterrupted sequence of channels between the probe and the goal. All that is required is to record, for each morphological feature, where this sequence ends (rewritability) and what path it takes (chain lineage). Thus, the two modifications introduced above are sufficient to keep track of channels connecting chains in an expression.

Example 6. Consider the expression exp, shown as a phrase-structure tree:

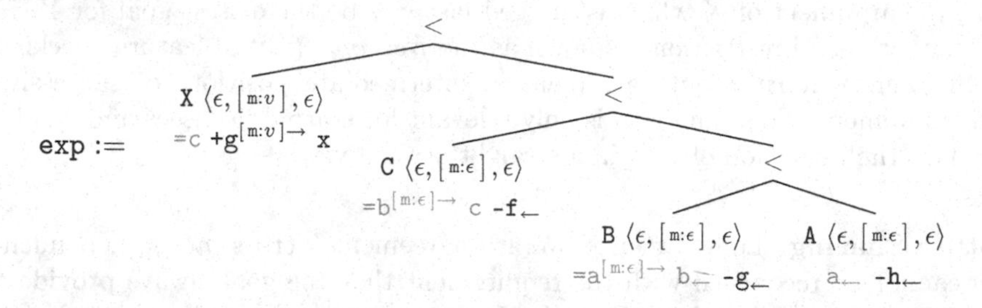

The next derivation step will engage the +g/-g feature pair, transmitting the value v to B; v has to percolate to A, but not to C. This information is lost in the standard chain notation. However, adding rewritabilities and lineages allows to identify chains accessible to agreement – namely, those with g in the lineage:

$$\begin{aligned}\texttt{exp}' := \ & \mathit{mrg3}\,(\texttt{X}, \mathit{mrg3}\,(\texttt{C}, \mathit{mrg3}\,(\texttt{B}, \texttt{A}))) = \\ & \langle \epsilon, [\,\texttt{m:}v/\texttt{off}\,], \epsilon \rangle : \texttt{+g}^{[\,\texttt{m:}v/\texttt{off}\,]\rightarrow}\ \texttt{x}\ (\texttt{off}), \\ & [\,\texttt{m:}\epsilon/\texttt{f}\,]\ \texttt{-f}_{\leftarrow}\ (\texttt{f off}),\ [\,\texttt{m:}\epsilon/\texttt{f}\,]\ \texttt{-g}_{\leftarrow}\ (\texttt{f g off}),\ [\,\texttt{m:}\epsilon/\texttt{f}\,]\ \texttt{-h}_{\leftarrow}\ (\texttt{f g h off})\end{aligned}$$

3.2 Minimalist Grammars with Agreement

Definition 7. A minimalist grammar with agreement (MG_{agr}) G is a 5-tuple $\langle Mor, Syn, Types, Lex, \mathcal{F}\rangle$, where

$Mor = \{f : X \rightarrow Base_m \times (\{\texttt{on}, \texttt{off}\} \cup Base)\}$ is a set of bundles, where $Base_m$, Val are finite sets such that $\epsilon \in Val$ is the default value, and $Base$ is a nonempty finite set (of syntactic feature names);

$Feat = Syn \times \{\leftarrow, \not\leftarrow\} \times Mor$ is a set of annotated features, where Syn is a set of syntactic features built from $Base$ as specified in Definition 2;

$Types = \{::, :\}$, (lexical, derived)

Let $Lineage = \{s \mid s \in Base^* \ \&\ \text{for } 1 \leq i, j \leq |s|,\ s_i \neq s_j\}\ \{\texttt{off}\}$. Then the set of initial chains $IC = Mor^* \times Mor^* \times Mor^*\ Types\ Feat^*\ Lineage$, and the set of non-initial chains $NC = Mor^*\ Feat^*\ Lineage$;

[5] If movement is viewed as copying, it is not immediately clear why this should be the case. A system where moving subtrees retain their relation to the original position would be interesting to explore but falls outside the scope of this paper.

$Lex \subset \{\epsilon\} \times Mor^* \times \{\epsilon\}\ \{::\}\ Feat^*\ \{\texttt{off}\}$, a subset of IC, is a finite set of lexical items,

$\mathcal{F} = \{\mathit{merge\text{-}agr}, \mathit{move\text{-}agr}\}$ is a set of structure-building operations.

Notation 8. Let $M \in Mor$ such that $M = \{\langle \phi_1, \langle v_1, r_1 \rangle\rangle, ..., \langle \phi_n, \langle v_n, r_n \rangle\rangle\}$. Then for $i \in [1..n]$, $v_i \in Val$ is the *value* of ϕ_i in M, and $r_i \in \{\texttt{on}, \texttt{off}\} \cup Base$ is its *rewritability*. M can be written as $\begin{bmatrix} \phi_1{:}v_1/r_1 \\ ... \\ \phi_n{:}v_n/r_n \end{bmatrix}$ or as $\varnothing$ if $n = 0$.

Notation 9. Let $f = \langle x\texttt{f},\ Y,\ M \rangle \in Feat$ such that $\texttt{f} \in Base$. Then $f_{\text{id}} = \texttt{f}$ is the name of f, $Y \in \{\leftarrow, \not\leftarrow\}$ specifies its receiving channel and $M \in Mor$ its emitting channel. f can be written as $x\texttt{f}_Y^{M\rightarrow}$. Where it does not lead to ambiguity, the receiving channel may be omitted if $Y = \not\leftarrow$, and the emitting channel may be omitted if $M = \varnothing$.

Notation 10. Chain lineages are enclosed in parentheses for better readability.

Notation 11. Let $\texttt{item} = \langle \epsilon, M, \epsilon \rangle\ ::\ \gamma\ (\texttt{off})$, where $M \in Mor$ and $\gamma \in Feat^*$, be a lexical item such that the bundle $M = \begin{bmatrix} \phi_1{:}v_1/r_1 \\ ... \\ \phi_n{:}v_n/r_n \end{bmatrix}$. Then $\begin{bmatrix} \text{ITEM} \\ \phi_{m+1}{:}v_{m+1}/r_{m+1} \\ ... \\ \phi_n{:}v_n/r_n \end{bmatrix}$ can be used as a semi-formal abbreviation for M, where ITEM stands for a subset of lexically valued features $\{\langle \phi_1, \langle v_1, \texttt{off} \rangle\rangle, ..., \langle \phi_m, \langle v_m, \texttt{off} \rangle\rangle\} \subseteq M$.

At the level of bundles, agreement is handled by two functions updating *active* (on) features with values provided by the goal. Downward agreement ($agr_\downarrow$) leaves all features in the probe on, as the probe is, by definition, part of the initial chain. Upward agreement ($agr_\uparrow$) sets each probe feature to on if active in the goal and to a given rewritability value otherwise. An auxiliary function, *act*, reactivates features with rewritability values present in a given sequence.

Definition 12. $agr_\downarrow : (Mor \times Mor) \rightarrow Mor$ is a function such that for $P, M \in Mor$, for $\phi \in Base_m$ the result of downward agreement of P with M is as follows:

$$agr_\downarrow(M, P, \phi) \equiv P^{\downarrow M}(\phi) = \begin{cases} \langle M(\phi)[1], \texttt{on} \rangle & \text{if } M(\phi) \text{ is defined} \\ & \text{and } P(\phi)[2] = \texttt{on}; \\ P(\phi) & \text{otherwise.} \end{cases}$$

Definition 13. $agr_\uparrow : ((\{\texttt{off}\} \cup Base) \times Mor \times Mor) \rightarrow Mor$ is a function such that for $P, M \in Mor$, $\texttt{re} \in \{\texttt{off}\} \cup Base$, $\phi \in Base_m$ the result of upward agreement of P with M (setting to $\texttt{re}$ the rewritability of any feature in P whose value is not to be shared with the feature in M) is as follows:

$$agr_\uparrow(\texttt{re}, M, P, \phi) \equiv P_{\texttt{re}}^{\uparrow M}(\phi) = \begin{cases} \langle M(\phi)[1], \texttt{on} \rangle & \text{if } P(\phi)[2] = \texttt{on} \\ & \text{and } M(\phi)[2] = \texttt{on}; \\ \langle M(\phi)[1], \texttt{re} \rangle & \text{if } P(\phi)[2] = \texttt{on} \\ & \text{and } M(\phi)[2] = \texttt{off}; \\ \langle P(\phi)[1], \texttt{re} \rangle & \text{if } P(\phi)[2] = \texttt{on} \\ & \text{and } M(\phi) \text{ is undefined;} \\ P(\phi) & \text{otherwise.} \end{cases}$$

Definition 14. $act : (Base^{+} \times Mor) \rightarrow Mor$ is a function such that for $P \in Mor$, $L \in Base^{+}$, $\phi \in Base_m$:

$$act(L, P, \phi) \equiv act_L(P, \phi) = \begin{cases} \langle P(\phi)[1], \mathtt{on} \rangle & \text{if } P(\phi)[2] \in L; \\ P(\phi) & \text{otherwise.} \end{cases}$$

The definitions are extended to apply to objects other than bundles:

Notation 15. For $M \in Mor$, $\mathtt{re} \in \{\mathtt{off}\} \cup Base$, $L \in Base^{+}$, $fun \in \{agr_{\downarrow}(M), agr_{\uparrow}(\mathtt{re}, M), act(L)\}$:

for $f \in Feat$, $fun(f) = \langle f[1], f[2], fun(f[3]) \rangle$;
for $(x_1, ..., x_n) \in Mor^{*}$, $fun(x_1, ..., x_n) = fun(x_1), ..., fun(x_n)$;
for $(x_1, ..., x_n) \in Feat^{*}$, $fun(x_1, ..., x_n) = fun(x_1), ..., fun(x_n)$;
for $c = s\, \gamma\, (A) \in NC$ such that $s \in Mor^{*}$ and $\gamma \in Feat^{*}$,
$fun(c) = fun(s)\, fun(\gamma)\, (A)$.

Finally, *merge* and *move* are redefined to accommodate agreement. The new rules manipulate lineages as well as bundles. Note that $P^{\downarrow\varnothing} = P$, but $P^{\uparrow\varnothing}_{\mathtt{re}} \neq P$: upward agreement with an empty bundle sets the rewritability of all features in the goal to $\mathtt{re}$. This special case corresponds to lack of agreement in the absence of a receiving channel and in intermediate positions of movement.

Definition 16. *merge-agr* is the union of the following five functions, for s_s, s_h, s_c, t_s, t_h, t_c, $t_1, ..., t_l \in Mor^{*}$, $\cdot \in \{:, ::\}$, $\mathtt{f}, \mathtt{x}_1, ..., \mathtt{x}_l \in Base$, $\gamma, \zeta \in Feat^{*}$, $\delta_1, ..., \delta_l \in Feat^{+}$, $M, N \in Mor$, $L_1, ..., L_l \in Lineage$, $X, Y \in \{\leftarrow, \not\leftarrow\}$, $\alpha_1, ..., \alpha_k$, $t_1\, \delta_1\, \mathtt{x}_1 L_1, ..., t_l\, \delta_l\, \mathtt{x}_l L_l \in NC$ $(0 \leq k, l)$,

mrg1-agr:

$$\frac{\langle \epsilon, s_h, \epsilon \rangle :: \mathtt{=f}^{M\rightarrow}_{X} \gamma\, (\mathtt{off}) \qquad \langle t_s, t_h, t_c \rangle \cdot \mathtt{f}^{N\rightarrow}_{Y}\, (\mathtt{off}),\, t_1\, \delta_1\, (\mathtt{x}_1 L_1), ..., t_l\, \delta_l\, (\mathtt{x}_l L_l)}{\langle \epsilon, s_h{}^{\downarrow\hat{N}}, (t_s t_h t_c)^{\uparrow\hat{M}}_{\mathtt{off}} \rangle : \gamma^{\downarrow\hat{N}}\, (\mathtt{off}),\, (t_1\, \delta_1\, (\mathtt{x}_1 L_1))^{\uparrow\hat{M}}_{\mathtt{x}_1}, ..., (t_l\, \delta_l\, (\mathtt{x}_l L_l))^{\uparrow\hat{M}}_{\mathtt{x}_l}}$$

mrg2-agr:

$$\frac{\langle s_s, s_h, s_c \rangle : \mathtt{=f}^{M\rightarrow}_{X} \gamma\, (\mathtt{off}),\, \alpha_1, ..., \alpha_k \qquad \langle t_s, t_h, t_c \rangle \cdot \mathtt{f}^{N\rightarrow}_{Y}\, [\mathtt{off}],\, t_1\, \delta_1\, (\mathtt{x}_1 L_1), ..., t_l\, \delta_l\, (\mathtt{x}_l L_l)}{\langle (t_s t_h t_c)^{\uparrow\hat{M}}_{\mathtt{off}} s_s{}^{\downarrow\hat{N}}, s_h{}^{\downarrow\hat{N}}, s_c{}^{\downarrow\hat{N}} \rangle : \gamma^{\downarrow\hat{N}}\, (\mathtt{off}),\, \alpha_1{}^{\downarrow\hat{N}}, ..., \alpha_k{}^{\downarrow\hat{N}},\, (t_1\, \delta_1\, (\mathtt{x}_1 L_1))^{\uparrow\hat{M}}_{\mathtt{x}_1}, ..., (t_l\, \delta_l\, (\mathtt{x}_l L_l))^{\uparrow\hat{M}}_{\mathtt{x}_l}}$$

mrg3-agr:

$$\frac{\langle s_s, s_h, s_c \rangle \cdot \mathtt{=f}^{M\rightarrow}_{X} \gamma\, (\mathtt{off}),\, \alpha_1, ..., \alpha_k \qquad \langle t_s, t_h, t_c \rangle \cdot \mathtt{f}^{N\rightarrow}_{Y} g\zeta\, (\mathtt{off}),\, t_1\, \delta_1\, (\mathtt{x}_1 L_1), ..., t_l\, \delta_l\, (\mathtt{x}_l L_l)}{\langle s_s{}^{\downarrow\hat{N}}, s_h{}^{\downarrow\hat{N}}, s_c{}^{\downarrow\hat{N}} \rangle : \gamma^{\downarrow\hat{N}}\, (\mathtt{off}),\, \alpha_1{}^{\downarrow\hat{N}}, ..., \alpha_k{}^{\downarrow\hat{N}},\, (t_s t_h t_c\, g\zeta\, (g_{\mathrm{id}}\mathtt{off}))^{\uparrow\hat{M}}_{g_{\mathrm{id}}},\, (t_1\, \delta_1\, (g_{\mathrm{id}} \mathtt{x}_1 L_1))^{\uparrow\hat{M}}_{g_{\mathrm{id}}}, ..., (t_l\, \delta_l\, (g_{\mathrm{id}} \mathtt{x}_l L_l))^{\uparrow\hat{M}}_{g_{\mathrm{id}}}}$$

hmrg1-agr:

$$\frac{\langle \epsilon, s_h, \epsilon \rangle :: \mathtt{=>f}^{M\rightarrow}_{X} \gamma\, (\mathtt{off}) \qquad \langle t_s, t_h, t_c \rangle \cdot \mathtt{f}^{N\rightarrow}_{Y}\, (\mathtt{off}),\, t_1\, \delta_1\, (\mathtt{x}_1 L_1), ..., t_l\, \delta_l\, (\mathtt{x}_l L_l)}{\langle \epsilon, t_h{}^{\uparrow\hat{M}}_{\mathtt{off}} s_h{}^{\downarrow\hat{N}}, (t_s t_c)^{\uparrow\hat{M}}_{\mathtt{off}} \rangle : \gamma^{\downarrow\hat{N}}\, (\mathtt{off}),\, (t_1\, \delta_1\, (\mathtt{x}_1 L_1))^{\uparrow\hat{M}}_{\mathtt{x}_1}, ..., (t_l\, \delta_l\, (\mathtt{x}_l L_l))^{\uparrow\hat{M}}_{\mathtt{x}_l}}$$

hmrg3-agr:

$$\frac{\langle s_s, s_h, s_c \rangle \cdot \mathtt{=>f}^{M\rightarrow}_{X} \gamma\, (\mathtt{off}),\, \alpha_1, ..., \alpha_k \qquad \langle t_s, t_h, t_c \rangle \cdot \mathtt{f}^{N\rightarrow}_{Y} g\zeta\, (\mathtt{off}),\, t_1\, \delta_1\, (\mathtt{x}_1 L_1), ..., t_l\, \delta_l\, (\mathtt{x}_l L_l)}{\langle s_s{}^{\downarrow\hat{N}}, t_h{}^{\uparrow\hat{M}}_{\mathtt{off}} s_h{}^{\downarrow\hat{N}}, s_c{}^{\downarrow\hat{N}} \rangle : \gamma^{\downarrow\hat{N}}\, (\mathtt{off}),\, \alpha_1{}^{\downarrow\hat{N}}, ..., \alpha_k{}^{\downarrow\hat{N}},\, (t_s t_c\, g\zeta\, (g_{\mathrm{id}}\mathtt{off}))^{\uparrow\hat{M}}_{g_{\mathrm{id}}},\, (t_1\, \delta_1\, (g_{\mathrm{id}} \mathtt{x}_1 L_1))^{\uparrow\hat{M}}_{g_{\mathrm{id}}}, ..., (t_l\, \delta_l\, (g_{\mathrm{id}} \mathtt{x}_l L_l))^{\uparrow\hat{M}}_{g_{\mathrm{id}}}}$$

$$\text{where } \hat{M} = \begin{cases} M & \text{if } Y = \leftarrow \\ \varnothing & \text{if } Y = \not\leftarrow \end{cases} \quad \text{and} \quad \hat{N} = \begin{cases} N & \text{if } X = \leftarrow \\ \varnothing & \text{if } X = \not\leftarrow \end{cases}$$

Movement rules have to address the additional complication of classifying chains. Each non-initial chain undergoes upward agreement (if it has the moving chain in its lineage) or downward agreement (as a subchain of the initial chain).

Definition 17. *move-agr* is the union of the following three functions, for s_s, s_h, $s_c, t_1, ..., t_k \in Mor^*$, $\mathtt{f} \in Base$, $\mathtt{F} \in \{\mathtt{-f}, \mathtt{*f}\}$, $\gamma, \zeta \in Feat^*$, $\delta_1, ..., \delta_k \in Feat^+$, $M, N \in Mor$, $A \in Base^*$, $B, L_1, ..., L_l \in Lineage$, $X, Y \in \{\leftarrow, \not\leftarrow\}$, and for $\alpha_1, ..., \alpha_k \in NC$ $(0 \leq k)$ such that for $j \in [1,k]$ $\alpha_j = t_j\ \delta_j\ L_j$, satisfying (SMC): there is exactly one $i \in [1,k]$ such that α_i has $\mathtt{-f}$ or $\mathtt{*f}$ as its first syntactic feature

mv1-agr:

$$\frac{\langle s_s, s_h, s_c \rangle \cdot \mathtt{+f}_X^{M\rightarrow} \gamma\,(\mathtt{off}),\ \alpha_1, ..., \alpha_{i-1},\ t_i\, \mathtt{F}_Y^{N\rightarrow}\ (A\mathtt{f}\ \mathtt{off}),\ \alpha_{i+1}, ..., \alpha_k}{\langle (act_{A\mathtt{f}}(t_i))_{\mathtt{off}}^{\uparrow\hat{M}} s_s{}^{\downarrow\hat{N}}, s_h{}^{\downarrow\hat{N}}, s_c{}^{\downarrow\hat{N}} \rangle\ :\ \gamma^{\downarrow\hat{N}}\,(\mathtt{off}),\ \alpha_1', ..., \alpha_{i-1}',\ \alpha_{i+1}', ..., \alpha_k'}$$

where for $j \in [1,k]$, $j \neq i$

$$\alpha_j' = \begin{cases} (act_{L_j'\mathtt{f}}(t_j\ \delta_j\ (\mathtt{x}L_j'')))_{\mathtt{x}}^{\uparrow\hat{M}} & \text{if } L_j = L_j'\mathtt{fx}L_j'' \text{ such that} \\ & L_j' \in Base^*, \mathtt{x} \in Base, L_j'' \in Lineage \\ (t_j\ \delta_j\ (L_j))^{\downarrow\hat{N}} & \text{otherwise} \end{cases}$$

mv2-agr:

$$\frac{\langle s_s, s_h, s_c \rangle \cdot \mathtt{+f}_X^{M\rightarrow} \gamma\,(\mathtt{off}),\ \alpha_1, ..., \alpha_{i-1},\ t_i\, \mathtt{F}_{\leftarrow}^{N\rightarrow} g\zeta\,(A\mathtt{f}B),\ \alpha_{i+1}, ..., \alpha_k}{\langle s_s{}^{\downarrow\hat{N}}, s_h{}^{\downarrow\hat{N}}, s_c{}^{\downarrow\hat{N}} \rangle\ :\ \gamma^{\downarrow\hat{N}}\,(\mathtt{off}),\ \alpha_1', ..., \alpha_{i-1}',\ (act_{A\mathtt{f}}(t_i\ g\zeta\ (g_{\mathrm{id}}B)))_{g_{\mathrm{id}}}^{\uparrow\hat{M}},\ \alpha_{i+1}', ..., \alpha_k'}$$

where for $j \in [1,k]$, $j \neq i$

$$\alpha_j' = \begin{cases} (act_{L_j'\mathtt{f}}(t_j\ \delta_j\ (g_{\mathrm{id}}L_j'')))_{g_{\mathrm{id}}}^{\uparrow\hat{M}} & \text{if } L_j = L_j'\mathtt{f}L_j'' \text{ such that} \\ & L_j' \in Base^*, L_j'' \in Lineage \\ (t_j\ \delta_j\ (L_j))^{\downarrow\hat{N}} & \text{otherwise} \end{cases}$$

mv-agr*:

$$\frac{\langle s_s, s_h, s_c \rangle \cdot \mathtt{+f}_X^{M\rightarrow} \gamma\,(\mathtt{off}),\ \alpha_1, ..., \alpha_{i-1},\ t_i\, \mathtt{*f}_Y^{N\rightarrow} \zeta\,(A\mathtt{f}B),\ \alpha_{i+1}, ..., \alpha_k}{\langle s_s, s_h, s_c \rangle\ :\ \gamma\,(\mathtt{off}),\ \alpha_1', ..., \alpha_{i-1}',\ (act_A(t_i\, \mathtt{*f}_Y^{N\rightarrow} \zeta\ (\mathtt{f}B)))_{\mathtt{f}}^{\uparrow\varnothing},\ \alpha_{i+1}', ..., \alpha_k'}$$

where for $j \in [1,k]$, $j \neq i$

$$\alpha_j' = \begin{cases} (act_{L_j'}(t_j\ \delta_j\ (\mathtt{f}L_j'')))_{\mathtt{f}}^{\uparrow\varnothing} & \text{if } L_j = L_j'\mathtt{f}L_j'' \text{ such that} \\ & L_j' \in Base^*, L_j'' \in Lineage \\ t_j\ \delta_j\ (L_j) & \text{otherwise} \end{cases}$$

$$\text{where } \hat{M} = \begin{cases} M & \text{if } Y = \leftarrow \\ \varnothing & \text{if } Y = \not\leftarrow \end{cases} \quad \text{and} \quad \hat{N} = \begin{cases} N & \text{if } X = \leftarrow \\ \varnothing & \text{if } X = \not\leftarrow \end{cases}$$

Example 18. $G' = \langle \Sigma_{G'}, Syn_{G'}, Types, Lex_{G'}, \mathcal{F} \rangle$ is a MG_{agr}. Its lexicon $Lex_{G'}$ contains the following lexical items:

$$\texttt{this} := \left\langle \epsilon, \begin{bmatrix} \text{THIS} \\ \texttt{num:}\epsilon\texttt{/on} \\ \texttt{per:}\epsilon\texttt{/on} \\ \texttt{case:}\epsilon\texttt{/on} \end{bmatrix}, \epsilon \right\rangle :: \texttt{=n}^{[\texttt{case:}\epsilon\texttt{/on}]\rightarrow}_{\leftarrow}\ \texttt{d -k}^{\begin{bmatrix} \texttt{num:}\epsilon\texttt{/on} \\ \texttt{per:}\epsilon\texttt{/on} \end{bmatrix}\rightarrow}_{\leftarrow}\ \texttt{(off)}$$

$$\texttt{boy.sg} := \left\langle \epsilon, \begin{bmatrix} \text{BOY} \\ \texttt{num:}\mathit{sg}\texttt{/off} \\ \texttt{per:}\mathit{3}\texttt{/off} \\ \texttt{case:}\epsilon\texttt{/on} \end{bmatrix}, \epsilon \right\rangle :: \texttt{n}^{\begin{bmatrix} \texttt{num:}\mathit{sg}\texttt{/off} \\ \texttt{per:}\mathit{3}\texttt{/off} \end{bmatrix}\rightarrow}_{\leftarrow}\ \texttt{(off)}$$

$$\texttt{walk} := \left\langle \epsilon, \begin{bmatrix} \text{WALK} \end{bmatrix}, \epsilon \right\rangle :: \texttt{=d v (off)}$$

$$\texttt{prs} := \left\langle \epsilon, \begin{bmatrix} \text{PRS} \\ \texttt{num:}\epsilon\texttt{/on} \\ \texttt{per:}\epsilon\texttt{/on} \end{bmatrix}, \epsilon \right\rangle :: \texttt{=v +k}^{[\texttt{case:}\mathit{nom}\texttt{/off}]\rightarrow}_{\leftarrow}\ \texttt{t (off)}$$

G' generates one expression of category $\texttt{t}$:

mv1-agr

$$\left\langle \begin{bmatrix} \text{THIS} \\ \texttt{num:}\mathit{sg}\texttt{/off} \\ \texttt{per:}\mathit{3}\texttt{/off} \\ \texttt{case:}\mathit{nom}\texttt{/off} \end{bmatrix} \begin{bmatrix} \text{BOY} \\ \texttt{num:}\mathit{sg}\texttt{/off} \\ \texttt{per:}\mathit{3}\texttt{/off} \\ \texttt{case:}\mathit{nom}\texttt{/off} \end{bmatrix}, \begin{bmatrix} \text{WALK} \end{bmatrix} \begin{bmatrix} \text{PRS} \\ \texttt{num:}\mathit{sg}\texttt{/on} \\ \texttt{per:}\mathit{3}\texttt{/on} \end{bmatrix}, \epsilon \right\rangle : \texttt{t (off)}$$

hmrg1-agr

$$\left\langle \epsilon, \begin{bmatrix} \text{WALK} \end{bmatrix} \begin{bmatrix} \text{PRS} \\ \texttt{num:}\epsilon\texttt{/on} \\ \texttt{per:}\epsilon\texttt{/on} \end{bmatrix}, \epsilon \right\rangle, \begin{bmatrix} \text{THIS} \\ \texttt{num:}\mathit{sg}\texttt{/k} \\ \texttt{per:}\mathit{3}\texttt{/k} \\ \texttt{case:}\epsilon\texttt{/k} \end{bmatrix} \begin{bmatrix} \text{BOY} \\ \texttt{num:}\mathit{sg}\texttt{/off} \\ \texttt{per:}\mathit{3}\texttt{/off} \\ \texttt{case:}\epsilon\texttt{/k} \end{bmatrix} : \texttt{+k}^{[\texttt{case:}\mathit{nom}\texttt{/off}]\rightarrow}_{\leftarrow}\ \texttt{t (off)}\ \texttt{-k}^{\begin{bmatrix} \texttt{num:}\mathit{sg}\texttt{/k} \\ \texttt{per:}\mathit{3}\texttt{/k} \end{bmatrix}\rightarrow}_{\leftarrow}\ \texttt{(k off)}$$

prs

$$\left\langle \epsilon, \begin{bmatrix} \text{PRS} \\ \texttt{num:}\epsilon\texttt{/on} \\ \texttt{per:}\epsilon\texttt{/on} \end{bmatrix}, \epsilon \right\rangle :: \texttt{=v +k}^{[\texttt{case:}\mathit{nom}\texttt{/off}]\rightarrow}_{\leftarrow}\ \texttt{t (off)}$$

mrg3-agr

$$\left\langle \epsilon, \begin{bmatrix} \text{WALK} \end{bmatrix}, \epsilon \right\rangle, \begin{bmatrix} \text{THIS} \\ \texttt{num:}\mathit{sg}\texttt{/k} \\ \texttt{per:}\mathit{3}\texttt{/k} \\ \texttt{case:}\epsilon\texttt{/k} \end{bmatrix} \begin{bmatrix} \text{BOY} \\ \texttt{num:}\mathit{sg}\texttt{/off} \\ \texttt{per:}\mathit{3}\texttt{/off} \\ \texttt{case:}\epsilon\texttt{/k} \end{bmatrix} : \texttt{v (off)}\ \texttt{-k}^{\begin{bmatrix} \texttt{num:}\mathit{sg}\texttt{/k} \\ \texttt{per:}\mathit{3}\texttt{/k} \end{bmatrix}\rightarrow}_{\leftarrow}\ \texttt{(k off)}$$

walk

$$\left\langle \epsilon, \begin{bmatrix} \text{WALK} \end{bmatrix}, \epsilon \right\rangle :: \texttt{=d v (off)}$$

mrg1-agr

$$\left\langle \epsilon, \begin{bmatrix} \text{THIS} \\ \texttt{num:}\mathit{sg}\texttt{/on} \\ \texttt{per:}\mathit{3}\texttt{/on} \\ \texttt{case:}\epsilon\texttt{/on} \end{bmatrix}, \begin{bmatrix} \text{BOY} \\ \texttt{num:}\mathit{sg}\texttt{/off} \\ \texttt{per:}\mathit{3}\texttt{/off} \\ \texttt{case:}\epsilon\texttt{/on} \end{bmatrix} \right\rangle : \texttt{d -k}^{\begin{bmatrix} \texttt{num:}\mathit{sg}\texttt{/on} \\ \texttt{per:}\mathit{3}\texttt{/on} \end{bmatrix}\rightarrow}_{\leftarrow}\ \texttt{(off)}$$

this

$$\left\langle \epsilon, \begin{bmatrix} \text{THIS} \\ \texttt{num:}\epsilon\texttt{/on} \\ \texttt{per:}\epsilon\texttt{/on} \\ \texttt{case:}\epsilon\texttt{/on} \end{bmatrix}, \epsilon \right\rangle :: \texttt{=n}^{[\texttt{case:}\epsilon\texttt{/on}]\rightarrow}_{\leftarrow}\ \texttt{d -k}^{\begin{bmatrix} \texttt{num:}\epsilon\texttt{/on} \\ \texttt{per:}\epsilon\texttt{/on} \end{bmatrix}\rightarrow}_{\leftarrow}\ \texttt{(off)}$$

boy.sg

$$\left\langle \epsilon, \begin{bmatrix} \text{BOY} \\ \texttt{num:}\mathit{sg}\texttt{/off} \\ \texttt{per:}\mathit{3}\texttt{/off} \\ \texttt{case:}\epsilon\texttt{/on} \end{bmatrix}, \epsilon \right\rangle :: \texttt{n}^{\begin{bmatrix} \texttt{num:}\mathit{sg}\texttt{/off} \\ \texttt{per:}\mathit{3}\texttt{/off} \end{bmatrix}\rightarrow}_{\leftarrow}\ \texttt{(off)}$$

4 Case Study: Dative Intervention in Icelandic

4.1 Data

A certain class of Icelandic constructions exhibits an interesting agreement pattern. The verb agrees in number with its nominative object inside a small clause (SC). However, this agreement seems optional: as an alternative, the verb may appear in the default 3.SG form. Furthermore, agreement can be disrupted by a dative experiencer intervening between the verb and the nominative object, in which case only the default verb form is possible. Only some experiencers cause this effect (1), while others are transparent for agreement (2):

(1) a. Það **finnst** fáum börnum [$_{SC}$ tölvurnar ljótar].
EXPL **find.SG** few children.DAT computers.DEF.NOM ugly.NOM

b. *Það **finnast** fáum börnum tölvurnar ljótar.
EXPL **find.PL** few children.DAT computers.DEF.NOM ugly.NOM

'Few children find the computers ugly.' [13, pp. 54–55]

(2) a. Það **finnst** mörgum stúdentum tölvurnar ljótar.
EXPL **find.SG** many students.DAT computers.DEF.NOM ugly.NOM

b. Það **finnast** mörgum stúdentum tölvurnar ljótar.
EXPL **find.PL** many students.DAT computers.DEF.NOM ugly.NOM

'Many students find the computers ugly.' [10, p. 1000]

The intervention effect can occur even if the dative undergoes wh-movement and no longer linearly intervenes between the verb and the nominative object:

(3) a. Hvaða stúdent **finnst** tölvurnar ljótar?
which student.DAT **find.SG** computers.DEF.NOM ugly.NOM

b. Hvaða stúdent ??**finnast** tölvurnar ljótar?
which student.DAT **find.PL** computers.DEF.NOM ugly.NOM

'Which student finds the computers ugly?' [10, p. 1001]

A generalization dealing with these examples is proposed in [13]: dative experiencers are transparent for agreement just in case they can undergo Object Shift – a movement to the specifier of *v*. The ability of a DP to shift is an independent property of the quantifier. Furthermore, [13] argues that agreement with the nominative is, in fact, deterministic: obligatory iff the experiencer *has* shifted, impossible otherwise. Object Shift is string-vacuous in examples like (2a). However, VP-level adverbs are expected to precede an in-situ dative and follow a shifted dative. The former configuration is compatible only with default agreement (4), while the latter only with normal agreement (5):

(4) a. Það **finnst** fljótt mörgum köttum mýsnar góðar.
EXPL **find.SG** quickly many cats.DEF.DAT mice.DEF.NOM tasty

b. Það ??/***finnast** fljótt mörgum köttum mýsnar góðar.
EXPL **find.PL** quickly many cats.DEF.DAT mice.DEF.NOM tasty

'Many cats quickly find the mice tasty.' [13, p. 63]

(5) a. Það ??/* **finnst** mörgum köttum fljótt mýsnar góðar.
EXPL **find.SG** many cats.DEF.DAT quickly mice.DEF.NOM tasty

b. Það **finnast** mörgum köttum fljótt mýsnar góðar.
EXPL **find.PL** many cats.DEF.DAT quickly mice.DEF.NOM tasty
'Many cats quickly find the mice tasty.' [13, p. 63]

This reasoning is extended to fronted dative experiencers, including wh-arguments (3), which are also required to undergo Object Shift for agreement to succeed. Importantly, Object Shift does not alter the relation between the T(ense) head, which morphologically manifests agreement, and the nominative object. This observation can be reconciled with the traditional notion of Agree by assuming that the primary locus of agreement is lower than T – namely, *v*. Object Shift removes the dative from the *probing domain* of *v*, allowing *v* to probe the nominative object. T inherits the relevant features from *v* (via Agree with other heads intervening between T and *v*). This approach to Agree reduces an instance of long-distance agreement to a series of local dependencies – not unlike the $\mathrm{MG}_{\mathrm{agr}}$ formalism.

4.2 Grammar Fragment

Let each determiner phrase transmit its number value via its category feature channel and the default value via its licensee channel. Then the difference between *shiftable* and *non-shiftable* DPs can be reduced to the distinction between a starred licensee (`*k`) and a plain licensee (`-k`).[6]

$$\texttt{many}\sim \quad := \left\langle \epsilon, \left[{\tiny\begin{matrix}\text{MANY}\sim\\ \texttt{num:}\mathit{pl}/\texttt{off}\end{matrix}}\right], \epsilon \right\rangle :: \texttt{d}^{[\texttt{num:}\mathit{pl}/\texttt{off}]\rightarrow}\ \texttt{*k}^{[\texttt{num:}\epsilon/\texttt{off}]\rightarrow} \quad (\texttt{off})$$

$$\texttt{few}\sim \quad := \left\langle \epsilon, \left[{\tiny\begin{matrix}\text{FEW}\sim\\ \texttt{num:}\mathit{pl}/\texttt{off}\end{matrix}}\right], \epsilon \right\rangle :: \texttt{d}^{[\texttt{num:}\mathit{pl}/\texttt{off}]\rightarrow}\ \texttt{-k}^{[\texttt{num:}\epsilon/\texttt{off}]\rightarrow} \quad (\texttt{off})$$

The verb `find` selects a small clause (containing an object DP) and an experiencer DP, receiving a number value from the former. There is no agreement with the dative experiencer, so the relevant selector `=d` has no receiving channel.

$$\texttt{find} \quad := \left\langle \epsilon, \left[\text{FIND}\right], \epsilon \right\rangle :: \texttt{=sc}_{\leftarrow}\ \texttt{=d}\ \texttt{V}^{[\texttt{num:}\epsilon/\texttt{on}]\rightarrow} \quad (\texttt{off})$$

$$\texttt{SC} \quad := \left\langle \epsilon, \left[{\tiny\begin{matrix}\text{SC}\\ \texttt{num:}\mathit{pl}/\texttt{off}\end{matrix}}\right], \epsilon \right\rangle :: \texttt{sc}^{[\texttt{num:}\mathit{pl}/\texttt{off}]\rightarrow} \quad (\texttt{off})$$

The crucial point in the derivation is `AgrO` which has two receiving channels. `v` and $\texttt{v}_{\texttt{shift}}$ pass information along; additionally, $\texttt{v}_{\texttt{shift}}$ provides the landing site of Object Shift. Thus, `T` eventually receives the last value transmitted to `AgrO`.

$$\texttt{AgrO} \quad := \left\langle \epsilon, \left[\text{AgrO}\right], \epsilon \right\rangle :: \texttt{=>V}_{\leftarrow}\ \texttt{+k}_{\leftarrow}\ \texttt{agrO}^{[\texttt{num:}\epsilon/\texttt{on}]\rightarrow} \quad (\texttt{off})$$

[6] For space reasons, DPs and small clauses are treated as atomic units. They can be decomposed to model internal agreement in detail. In particular, it is possible to embed DPs within another functional projection, connecting intervention and case (cf. [14], i.a.); this would allow the verb to assign dative to its argument and manipulate the agreement properties of its outer layer at the same time.

$$\texttt{v} \quad := \left\langle \epsilon, [\,\textsc{v}\,], \epsilon \right\rangle :: \texttt{=>agrO}_{\leftarrow}\ \texttt{v}^{[\texttt{num}:\epsilon/\texttt{on}]\rightarrow}\ (\texttt{off})$$

$$\texttt{v}_{\texttt{shift}} \quad := \left\langle \epsilon, [\,\textsc{v}_{\textsc{shift}}\,], \epsilon \right\rangle :: \texttt{=>agrO}_{\leftarrow}\ \texttt{+k}\ \texttt{v}^{[\texttt{num}:\epsilon/\texttt{on}]\rightarrow}\ (\texttt{off})$$

$$\texttt{T} \quad := \left\langle \epsilon, \left[\begin{smallmatrix}\text{T}\\ \texttt{num}:\epsilon/\texttt{on}\end{smallmatrix}\right], \epsilon \right\rangle :: \texttt{=>v}_{\leftarrow}\ \texttt{t}\ \ (\texttt{off})$$

Abstracting away from person and case, the eight lexical items defined above suffice[7] to model the number contrast between (1) and (2). Any experiencer can check its licensee in the specifier of AgrO. In this case, AgrO receives the default number value, giving rise to (1a) and (2a).

Example 19. The derivation of (1a) proceeds as follows:

$$\texttt{i1} := \textit{mrg1-agr}\,(\texttt{find}, \texttt{sc})$$
$$= \left\langle \epsilon, [\,\textsc{find}\,], \left[\begin{smallmatrix}\text{SC}\\ \texttt{num}:pl/\texttt{off}\end{smallmatrix}\right] \right\rangle : \texttt{=d}\,\texttt{V}^{[\texttt{num}:pl/\texttt{on}]\rightarrow}\,(\texttt{off})$$

$$\texttt{i2} := \textit{mrg3-agr}\,(\texttt{i1}, \texttt{few}\sim)$$
$$= \left\langle \epsilon, [\,\textsc{find}\,], \left[\begin{smallmatrix}\text{SC}\\ \texttt{num}:pl/\texttt{off}\end{smallmatrix}\right] \right\rangle : \texttt{V}^{[\texttt{num}:pl/\texttt{on}]\rightarrow}\,(\texttt{off}),$$
$$\left[\begin{smallmatrix}\text{FEW}\sim\\ \texttt{num}:pl/\texttt{off}\end{smallmatrix}\right] \texttt{-k}^{[\texttt{num}:\epsilon/\texttt{off}]\rightarrow}\,(\texttt{off})$$

$$\texttt{i3} := \textit{hmrg1-agr}\,(\texttt{AgrO}, \texttt{i2})$$
$$= \left\langle \epsilon, [\,\textsc{find}\,][\,\textsc{AgrO}\,], \left[\begin{smallmatrix}\text{SC}\\ \texttt{num}:pl/\texttt{off}\end{smallmatrix}\right] \right\rangle :: \texttt{+k}_{\leftarrow}\,\texttt{agrO}^{[\texttt{num}:pl/\texttt{on}]\rightarrow}\,(\texttt{off}),$$
$$\left[\begin{smallmatrix}\text{FEW}\sim\\ \texttt{num}:pl/\texttt{off}\end{smallmatrix}\right] \texttt{-k}^{[\texttt{num}:\epsilon/\texttt{off}]\rightarrow}\,(\texttt{k off})$$

$$\texttt{i4} := \textit{mv1-agr}\,(\texttt{i3})$$
$$= \left\langle \left[\begin{smallmatrix}\text{FEW}\sim\\ \texttt{num}:pl/\texttt{off}\end{smallmatrix}\right], [\,\textsc{find}\,][\,\textsc{AgrO}\,], \left[\begin{smallmatrix}\text{SC}\\ \texttt{num}:pl/\texttt{off}\end{smallmatrix}\right] \right\rangle : \texttt{agrO}^{[\texttt{num}:\epsilon/\texttt{on}]\rightarrow}\,(\texttt{off})$$

$$\texttt{i5} := \textit{hmrg1-agr}\,(\texttt{v}, \texttt{i4})$$
$$= \left\langle \epsilon, [\,\textsc{find}\,][\,\textsc{AgrO}\,][\,\textsc{v}\,], \left[\begin{smallmatrix}\text{FEW}\sim\\ \texttt{num}:pl/\texttt{off}\end{smallmatrix}\right]\left[\begin{smallmatrix}\text{SC}\\ \texttt{num}:pl/\texttt{off}\end{smallmatrix}\right] \right\rangle : \texttt{v}^{[\texttt{num}:\epsilon/\texttt{on}]\rightarrow}\,(\texttt{off})$$

$$\texttt{i6} := \textit{hmrg1-agr}\,(\texttt{T}, \texttt{i5})$$
$$= \left\langle \epsilon, [\,\textsc{find}\,][\,\textsc{AgrO}\,][\,\textsc{v}\,]\left[\begin{smallmatrix}\text{T}\\ \texttt{num}:\epsilon/\texttt{on}\end{smallmatrix}\right], \left[\begin{smallmatrix}\text{FEW}\sim\\ \texttt{num}:pl/\texttt{off}\end{smallmatrix}\right]\left[\begin{smallmatrix}\text{SC}\\ \texttt{num}:pl/\texttt{off}\end{smallmatrix}\right] \right\rangle : \texttt{t}\,(\texttt{off})$$

[7] These items support partial derivations up to T, where morphological dependencies are resolved. Assuming that expletives are merged above TP and move to the specifier of CP [3], the following addition enables full CP derivations of (1)–(3):

$$\texttt{which}\sim \quad := \left\langle \epsilon, \left[\begin{smallmatrix}\text{WHICH}\sim\\ \texttt{num}:pl/\texttt{off}\end{smallmatrix}\right], \epsilon \right\rangle :: \texttt{d}^{[\texttt{num}:pl/\texttt{off}]\rightarrow}\ \texttt{-k}^{[\texttt{num}:\epsilon/\texttt{off}]\rightarrow}\ \texttt{-wh}\ \ (\texttt{off})$$

$$\texttt{T}_{\texttt{expl}} \quad := \left\langle \epsilon, [\,\textsc{T}_{\textsc{expl}}\,], \epsilon \right\rangle :: \texttt{=t =expl t}\ \ (\texttt{off})$$

$$\texttt{Expl} \quad := \left\langle \epsilon, [\,\textsc{expl}\,], \epsilon \right\rangle :: \texttt{expl -wh}\ (\texttt{off})$$

$$\texttt{C} \quad := \left\langle \epsilon, [\,\textsc{c}\,], \epsilon \right\rangle :: \texttt{=t +wh c}\ \ (\texttt{off})$$

The second option is only available for shiftable experiencers. The $*\texttt{k}$ feature can "survive" movement to the specifier of $\texttt{AgrO}$ to be checked later, in the specifier of $\texttt{v}_{\texttt{shift}}$ (whose $\texttt{+k}$ has no receiving channel): this movement corresponds to Object Shift. In this case, $\texttt{AgrO}$ will never receive the default agreement value and will instead transmit whatever value came from the small clause, resulting in the verb agreeing with its nominative argument.

Example 20. The derivation of (2b) proceeds as follows:

$$\texttt{a1} := \mathit{mrg1\text{-}agr}\,(\texttt{find}, \texttt{sc})$$

$$= \left\langle \epsilon, \left[\,\textsc{find}\,\right], \left[\begin{array}{c}\text{SC}\\ \texttt{num:}\mathit{pl}\texttt{/off}\end{array}\right] \right\rangle \;:\; \texttt{=d}\,\texttt{V}^{[\texttt{num:}\mathit{pl}\texttt{/on}]\rightarrow}\,(\texttt{off})$$

$$\texttt{a2} := \mathit{mrg3\text{-}agr}\,(\texttt{a1}, \texttt{many}{\sim})$$

$$= \left\langle \epsilon, \left[\,\textsc{find}\,\right], \left[\begin{array}{c}\text{SC}\\ \texttt{num:}\mathit{pl}\texttt{/off}\end{array}\right] \right\rangle \;:\; \texttt{V}^{[\texttt{num:}\mathit{pl}\texttt{/on}]\rightarrow}\,(\texttt{off}),$$

$$\left[\begin{array}{c}\textsc{many}{\sim}\\ \texttt{num:}\mathit{pl}\texttt{/off}\end{array}\right] *\texttt{k}^{[\texttt{num:}\epsilon\texttt{/off}]\rightarrow}\,(\texttt{k off})$$

$$\texttt{a3} := \mathit{hmrg1\text{-}agr}\,(\texttt{AgrO}, \texttt{a2})$$

$$= \left\langle \epsilon, \left[\,\textsc{find}\,\right]\left[\,\textsc{AgrO}\,\right], \left[\begin{array}{c}\text{SC}\\ \texttt{num:}\mathit{pl}\texttt{/off}\end{array}\right] \right\rangle \;:\; \texttt{+k}_{\leftarrow}\,\texttt{agrO}^{[\texttt{num:}\mathit{pl}\texttt{/on}]\rightarrow}\,(\texttt{off}),$$

$$\left[\begin{array}{c}\textsc{many}{\sim}\\ \texttt{num:}\mathit{pl}\texttt{/off}\end{array}\right] *\texttt{k}^{[\texttt{num:}\epsilon\texttt{/off}]\rightarrow}\,(\texttt{k off})$$

$$\texttt{a4} := \mathit{mv{*}\text{-}agr}\,(\texttt{a3})$$

$$= \left\langle \epsilon, \left[\,\textsc{find}\,\right]\left[\,\textsc{AgrO}\,\right], \left[\begin{array}{c}\text{SC}\\ \texttt{num:}\mathit{pl}\texttt{/off}\end{array}\right] \right\rangle \;:\; \texttt{agrO}^{[\texttt{num:}\mathit{pl}\texttt{/on}]\rightarrow}\,(\texttt{off}),$$

$$\left[\begin{array}{c}\textsc{many}{\sim}\\ \texttt{num:}\mathit{pl}\texttt{/off}\end{array}\right] *\texttt{k}^{[\texttt{num:}\epsilon\texttt{/off}]\rightarrow}\,(\texttt{k off})$$

$$\texttt{a5} := \mathit{hmrg1\text{-}agr}\,(\texttt{v}_{\texttt{shift}}, \texttt{a4})$$

$$= \left\langle \epsilon, \left[\,\textsc{find}\,\right]\left[\,\textsc{AgrO}\,\right]\left[\,\textsc{v}_{\textsc{shift}}\,\right], \left[\begin{array}{c}\text{SC}\\ \texttt{num:}\mathit{pl}\texttt{/off}\end{array}\right] \right\rangle \;:\; \texttt{+k}\,\texttt{v}^{[\texttt{num:}\mathit{pl}\texttt{/on}]\rightarrow}\,(\texttt{off}),$$

$$\left[\begin{array}{c}\textsc{many}{\sim}\\ \texttt{num:}\mathit{pl}\texttt{/off}\end{array}\right] *\texttt{k}^{[\texttt{num:}\epsilon\texttt{/off}]\rightarrow}\,(\texttt{k off})$$

$$\texttt{a6} := \mathit{mv1\text{-}agr}\,(\texttt{a5})$$

$$= \left\langle \left[\begin{array}{c}\textsc{many}{\sim}\\ \texttt{num:}\mathit{pl}\texttt{/off}\end{array}\right], \left[\,\textsc{find}\,\right]\left[\,\textsc{AgrO}\,\right]\left[\,\textsc{v}_{\textsc{shift}}\,\right], \left[\begin{array}{c}\text{SC}\\ \texttt{num:}\mathit{pl}\texttt{/off}\end{array}\right] \right\rangle \;:\; \texttt{v}^{[\texttt{num:}\mathit{pl}\texttt{/on}]\rightarrow}\,(\texttt{off})$$

$$\texttt{a7} := \mathit{hmrg1\text{-}agr}\,(\texttt{T}, \texttt{a6})$$

$$= \left\langle \epsilon, \left[\,\textsc{find}\,\right]\left[\,\textsc{AgrO}\,\right]\left[\,\textsc{v}_{\textsc{shift}}\,\right]\left[\begin{array}{c}\text{T}\\ \texttt{num:}\mathit{pl}\texttt{/on}\end{array}\right], \left[\begin{array}{c}\textsc{many}{\sim}\\ \texttt{num:}\mathit{pl}\texttt{/off}\end{array}\right]\left[\begin{array}{c}\text{SC}\\ \texttt{num:}\mathit{pl}\texttt{/off}\end{array}\right] \right\rangle \;:\; \texttt{t}\,(\texttt{off})$$

5 Discussion

In this paper, I have developed a modification of Minimalist Grammars that accommodates morphological agreement, redefining syntactic operations over bundles of morphological features. The Agree operation of Chomsky's Minimalist Program is reduced to local transmission of information over syntactic dependencies. In order to demonstrate the practicality of the new formalism, I have used it to express a precise, formalized analysis of Icelandic dative intervention inspired by the proposal in [13]. The key element of the generalization – namely, the link between Object Shift and agreement – has been translated into a grammar fragment, which can be expanded further to incorporate more insights from Minimalist syntax (see e.g. fn. 6, 7).

MGs with agreement output sequences of bundles and can interface with any sufficiently explicit theory of morphology. For instance, they are compatible with formalizations of Distributed Morphology [9] that take a sequence of feature structures as syntactic input: [19] and, more recently, [6].

While proving the equivalence of MGs with agreement and unmodified MGs is outside the scope of this paper, a few observations on the matter are in order. Agreement is transmitted across dependencies established by structure-building operations. This limits the number of goals in any given expression to the number of chains, which, in turn, is guaranteed to be finite by the SMC. The problem of long-distance upward agreement, which is the only remaining source of "nonlocality" in the grammar, is addressed by adopting an approach similar to feature sharing: informally speaking, each item that requires a feature value via upward agreement pushes the responsibility for obtaining it to whatever item immediately selects or licenses it. Thus, at any given point in the derivation, there are finitely many *different* morphological feature values which can be updated or transmitted to other items. This makes it possible to convert an MG_{agr} into an equivalent MG, reformulating lexical items over bundles in terms of unanalyzable elements (corresponding to bundles of valued features) and recasting agreement transformations as compatibility constraints.

Acknowledgments. My thanks to Karlos Arregi and Greg Kobele for their advice regarding both the linguistic and the formal aspects of this paper, and to the anonymous reviewers for their helpful comments and suggestions.

References

1. Adger, D.: A minimalist theory of feature structure. In: Features: Perspectives on a Key Notion in Linguistics, pp. 185–218 (2010)
2. Bošković, Ž.: On successive cyclic movement and the freezing effect of feature checking. In: Hartmann, J., Hegedüs, V., van Riemsdijk, H. (eds.) Sounds of Silence: Empty Elements in Syntax and Phonology. Elsevier North Holland, Amsterdam (in press)
3. Bowers, J.: Transitivity. Linguist. Inquiry **33**(2), 183–224 (2002)
4. Chomsky, N.: The Minimalist Program. MIT Press, Cambridge (1995)
5. Chomsky, N.: Minimalist inquiries: the framework. In: Martin, R., Michaels, D., Uriagereka, J. (eds.) Step by Step: Essays on Minimalist Syntax in Honor of Howard Lasnik, pp. 89–156. MIT Press, Cambridge (2000)
6. Ermolaeva, M., Edmiston, D.: Distributed morphology over strings. Poster presented at PLC 41, Philadelphia (2017)
7. Frampton, J., Gutmann, S.: Agreement is Feature Sharing (2000, unpublished manuscript)
8. Graf, T.: Local and transderivational constraints in syntax and semantics. Ph.D. thesis, UCLA (2013)

9. Halle, M., Marantz, A.: Distributed morphology and the pieces of inflection. In: Hale, K., Keyser, S. (eds.) The View from Building 20, pp. 111–176. The MIT Press, Cambridge (1993)
10. Holmberg, A., Hróarsdóttir, T.: Agreement and movement in Icelandic raising constructions. Lingua **114**(5), 651–673 (2004)
11. Kobele, G.M.: Generating copies: an investigation into structural identity in language and grammar. Ph.D. thesis, UCLA (2006)
12. Kobele, G.M.: A derivational approach to phrasal spellout. Slides presented at BCGL 7 (2012)
13. Kučerová, I.: Long-distance agreement in Icelandic: locality restored. J. Comp. German. Linguist. **19**(1), 49–74 (2016)
14. Rezac, M.: Phi-agree and theta-related case. In: Harbour, D., Adger, D., Bejar, S. (eds.) Phi Theory: Phi Features across Modules and Interfaces, pp. 83–130. The MIT Press, Cambridge (2008)
15. Shieber, S.M.: An Introduction to Unification-based Approaches to Grammar. CSLI Lecture Notes. CSLI, Stanford (1986)
16. Stabler, E.: Derivational minimalism. In: Retoré, C. (ed.) LACL 1996. LNCS, vol. 1328, pp. 68–95. Springer, Heidelberg (1997). https://doi.org/10.1007/BFb0052152
17. Stabler, E.P.: Recognizing head movement. In: de Groote, P., Morrill, G., Retoré, C. (eds.) LACL 2001. LNCS (LNAI), vol. 2099, pp. 245–260. Springer, Heidelberg (2001). https://doi.org/10.1007/3-540-48199-0_15
18. Stabler, E.P.: Computational perspectives on minimalism. In: Boeckx, C. (ed.) Oxford Handbook of Linguistic Minimalism, pp. 617–643. Oxford University Press, Oxford (2011)
19. Trommer, J.: Morphology consuming syntax' resources: generation and parsing in a minimalist version of distributed morphology. In: Proceedings of the ESSLI Workshop on Resource Logics and Minimalist Grammars (1999)

A Model-Theoretic Reconstruction of Type-Theoretic Semantics for Anaphora

Matthew Gotham[(✉)]

University of Oslo, Oslo, Norway
matthew.gotham@ifikk.uio.no

Abstract. I present an analysis of the interpretation of anaphora that takes concepts from type-theoretic semantics, in particular the use of the Σ and Π dependent type constructors, and incorporates them into a model-theoretic framework. The analysis makes use of (parametrically) polymorphic lexical entries. The key ideas are that, in the simplest case, eventualities can play the role that proof objects do in type-theoretic semantics; that more complex, compositionally-defined, structures can play that role in other cases; and that pronouns can be modelled by context-dependent functions from proof objects of the preceding discourse (in this sense) to entities.

Keywords: Anaphora · Donkey sentences · Polymorphism

1 Introduction

Type-theoretic semantics (TTS) is a variety of proof-theoretic semantics according to which the meaning of a sentence is a type in some underlying type theory, such that objects of that type are proofs of the proposition expressed by the sentence; we may say, then, that the sentence is true if and only if there is some object of that type [18].

Type theories chosen for TTS tend to be based on the intuitionistic type theory (ITT) of Martin-Löf [15], which contributes the particularly important (for TTS analyses) dependent type constructors Σ and Π, defined as in (1).

(1) a. If A is a type and, on the assumption that x is of type A, B is a type, then $(\Sigma x : A)(B)$ and $(\Pi x : A)(B)$ are types.
b. Objects of type $(\Sigma x : A)(B)$ are ordered pairs $\langle a, b\rangle$ such that a is of type A and b is of type $B[a/x]$.
c. Objects of type $(\Pi x : A)(B)$ are functions f with domain A such that for any object a of type A, $f(a)$ is of type $B[a/x]$.

In TTS, Σ and Π types can be used to give the meanings of existentially and universally quantified sentences, respectively.[1] For example, on the assumptions

[1] Σ can also be used to give the meaning of conjunction, and Π implication; this is reflected in the lexical entries given in Fig. 1. Limitations of space prevent any further consideration of these connections here.

A. Foret et al. (Eds.): FG 2017, LNCS 10686, pp. 37–53, 2018.
https://doi.org/10.1007/978-3-662-56343-4_3

that DONKEY is the type of donkeys and that (for any x) $\text{BRAY}(x)$ is the type of proofs that x is braying, (2-a)–(2-b) have the interpretations given by the types shown in (3-a)–(3-b) respectively.

(2) a. A donkey is braying.
 b. Every donkey is braying.

(3) a. $(\Sigma x : \text{DONKEY})(\text{BRAY}(x))$
 b. $(\Pi x : \text{DONKEY})(\text{BRAY}(x))$

In this analysis, (2-a) is true iff there is some object of the type shown in (3-a), i.e. an ordered pair consisting of a donkey and a proof that that donkey is braying. Likewise, (2-b) is true iff there is some object of the type shown in (3-b), i.e. a function mapping every donkey to a proof that that donkey is braying.

TTS gives us the resources to formalize the famous 'donkey sentence' (4) from [6], in a way that respects the syntax of the English sentence.

(4) Every farmer who owns a donkey beats it.

On the most natural interpretation of (4), the interpretation of *it* co-varies with that of *a donkey*. The well-known problem that this example poses for model-theoretic semantic (MTS) theories in the tradition of [16] is that this co-variation cannot straightforwardly be accounted for. In the natural naïve formalization of (4), shown in (5), the variable y in the consequent of the conditional is outside the scope of the existential quantifier.[2] Its interpretation would therefore not covary with donkeys according to standard model theory.[3]

(5) $\forall x.(\mathsf{farmer}(x) \wedge \exists y.\mathsf{donkey}(y) \wedge \mathsf{own}(x, y)) \rightarrow \mathsf{beat}(x, y)$

There is a truth-conditionally adequate formalization of (4), shown in (6), but this leaves an explanatory gap as to why the (normally) existential *a donkey* should end up being translated as a universal quantifier.

(6) $\forall x.\forall y.(\mathsf{farmer}(x) \wedge (\mathsf{donkey}(y) \wedge \mathsf{own}(x, y))) \rightarrow \mathsf{beat}(x, y)$

By contrast, as first pointed out in [21], in TTS the meaning of (4) can be expressed by the type shown in (7), where Π has been given as the meaning of *every* and Σ has been given as the meaning of *a*, as expected. (This translation also makes use of the projections p and q, where for any ordered pair $\langle a, b\rangle$, $p(\langle a, b\rangle) = a$ and $q(\langle a, b\rangle) = b$.)

(7) $(\Pi z : (\Sigma x : \text{FARMER})((\Sigma y : \text{DONKEY})(\text{OWN}(x, y))))(\text{BEAT}(p(z), p(q(z))))$

[2] Here and throughout the paper, a dot following a variable binder will often be used instead of parentheses to indicate unbounded scope to the right.

[3] One option, therefore, is to change the model theory so that (5) would be interpreted in the desired way. This, explicitly, is the approach taken in Dynamic Predicate Logic (DPL) [8]. Discourse Representation Theory (DRT) [13] and File Change Semantics (FCS) [10] take a similar approach.

The type shown in (7) is the type of functions f such that:

- the domain of f is the set of ordered pairs $\langle a, b\rangle$ such that:
  - a is a farmer, and
  - b is an ordered pair $\langle c, d\rangle$ such that
    - c is a donkey, and
    - d is a proof that a owns c, and
- f maps every $\langle a, \langle c, d\rangle\rangle$ in its domain to a proof that a beats c.

If and only if there is such a function, then there is an object of the type given in (7), and therefore, according to TTS, (4) is true on the intended interpretation.

The insight that ITT can be fruitfully applied to natural language semantics has been expanded into a detailed system in [18], and the analysis of anaphora, including especially cases like (4), has been developed and improved recently in [2]. In this paper, I will give an implementation of the ideas underlying the TTS analysis of anaphora, largely based on [2], in a model-theoretic framework.[4] This implementation has a two-fold motivation. Firstly, it enables us to examine the extent to which the TTS account of anaphora genuinely depends on an enriched type theory. Secondly, it provides insight into the relationship between the TTS account of anaphora and some 'dynamic' accounts in the MTS literature, given the similarity between the system that we end up with and some of them.

The paper is structured as follows. In Sect. 2 a translation of the account of [2] into higher-order logic will be built up in stages. In Sect. 3 I will expand the analysis to generalized quantifiers and plurals. Section 4 concludes.

2 Translating TTS for Anaphora into MTS

2.1 A First Pass

As a first step in our implementation we can consider again the interpretations given by TTS to (2-a)–(2-b) shown in (3-a)–(3-b) respectively. For the case of the Σ type constructor, and hence the existential claim, we have the gloss shown in (8).

(8) *A donkey is braying* is true iff there is an ordered pair consisting of a donkey and a proof that donkey is braying.

What would constitute a proof that a donkey is braying? In the light of the helpful discussion in [18], Sect. 2.26, I will take the proof object to be an event, and in general the proof objects denoted by VPs to be eventualities. We will need in our underlying type theory a basic type v for eventualities, then, in addition to the basic types e for entities and t for truth values. For reasons to be discussed in Sect. 2.2, I will also assume that the unit type 1 is among our basic

[4] By this, I mean that meanings will be given as expressions of a logical language, which are taken to be dispensable in favour of *their* interpretations in a model (as in [16]), which is where the 'real' semantics is. Expressions of the language of type theory are not understood this way in TTS; see [14] and [18], Sect. 2.27.

types. As (8) shows that we want to express ordered pairs, we will also need the type constructor $\times$ for binary product types in addition to the standard type constructor $\rightarrow$ for functional types, along with associated term constructors $(.,.)$ for pairing and $[.]_0$ and $[.]_1$ for left and right projections, respectively.

Given these considerations, (2-a) can be translated into higher-order logic as shown in (9). Note that I am assuming a 'Davidsonian' approach to events according to which predicates denote relations between individuals and events directly, rather than mediated by theta roles.

(9) $\exists a^{e \times v}.\mathsf{donkey}([a]_0) \wedge \mathsf{bray}(a)$
$\equiv$ $\exists x^e.\exists e^v.\mathsf{donkey}(x) \wedge \mathsf{bray}(x, e)$

However, recall that the non-empty type condition reflected in (8) ('there is...') comes from TTS—that is to say, from the natural language semantic application of ITT and not from the definitions of the type constructors themselves; it is not, for example, reflected in (3-a). It therefore seems more appropriate to say that the compositionally-constructed interpretation of (2-a) is as shown in (10), and that existential closure of the abstracted variable a comes about from a discourse process. As we will see in Sect. 2.2, this will also allow indefinites to bind pronouns outside of what is normally thought to be their scope.

(10) $\lambda a^{e \times v}.\mathsf{donkey}([a]_0) \wedge \mathsf{bray}(a)$

These considerations lead to the provisional lexical entry for a, expressed in TTS by Σ, shown in (11).[5]

(11) $\lambda P^{e \rightarrow t}.\lambda V^{e \rightarrow v \rightarrow t}.\lambda a^{e \times v}.P([a]_0) \wedge V([a]_0)([a]_1)$

Now let us consider the Π type constructor, and the interpretation given to (2-b) in (3-b) as glossed in (12).

(12) *Every donkey is braying* is true iff there is a function mapping every donkey to a proof that donkey is braying.

Here, we meet a complication that is not present in the discussion of Σ. Given the definition of the Π type constructor, the domain of the function alluded to in (12) should just be the set of donkeys. But this is not straightforward to accomplish in higher-order logic without having a type of donkeys—which is precisely a feature of TTS that we want to eliminate. The technique that I will adopt at this point is simply to say that the function is defined on the whole domain of entities, but that its interpretation is constrained in the right way whenever it applies to a donkey. Therefore, (3-b) can be translated as shown in (13), and hence the provisional lexical entry for *every*, expressed in TTS by Π, can be given as shown in (14).

[5] In the type annotations, here and throughout the rest of the paper, brackets are omitted where possible, on the understanding that both $\times$ and $\rightarrow$ associate to the right and that $\times$ binds more tightly than $\rightarrow$.

(13) $\lambda f^{e \to v}.\forall x^e.\mathsf{donkey}(x) \to \mathsf{bray}(x, f(x))$

(14) $\lambda P^{e \to t}.\lambda V^{e \to v \to t}.\lambda f^{e \to v}.\forall x^e.P(x) \to V(x)(f(x))$

We meet another complication when we consider the interpretation of the relative pronoun *who*. As shown in (7), the TTS analysis formalizes this using Σ, like the indefinite article—this, in fact, is part of what makes the extended binding scope possible. So we might expect the lexical entry to be as shown in (15).

(15) $\lambda V^{e \to v \to t}.\lambda P^{e \to t}.\lambda a^{e \times v}.P([a]_0) \wedge V([a]_0)([a]_1)$

However, on the basis of the provisional lexical entry given for *a* in (11) and other natural assumptions, the VP *owns a donkey* would be translated as shown in (16), and the types of (15) and (16) don't fit together.

(16) $\lambda x^e.\lambda a^{e \times v}.\mathsf{donkey}([a]_0) \wedge \mathsf{own}(x, a)$

This example shows up the need for some limited polymorphism in our lexical entries. (16) is not quite of the right type to be an argument for (15) because it contains extra information about an indefinite that was an argument to the embedded verb, and therefore is of type $e \to e \times v \to t$ rather than $e \to v \to t$ as expected by (15).

A similar issue is brought to light if we use (15) to derive the interpretation of a modified noun not containing any indefinites, for example *donkey who brays*, as shown in (17).

(17) $\lambda a^{e \times v}.\mathsf{donkey}([a]_0) \wedge \mathsf{bray}(a)$

Unsurprisingly—given that (15) is just a permutation of (11)—(17) is identical to the interpretation derived for *a donkey brays* in (10). It is of type $e \times v \to t$, not $e \to t$, and therefore not the right type to be an argument to (11).

The final limitation of the account so far is that there is no obvious way to incorporate pronouns. In the system set out in the next section, I will follow [2] and address this limitation by introducing a notion of context and a mechanism for updating it.

2.2 Full Implementation

In Fig. 1 I have given an initial list of lexical entries for a fragment to be used in this paper.

Some remarks are in order. Firstly, how should we understand the lowercase Greek letters in the type annotations? I prefer to think of them as metavariables over types, such that what we have in Fig. 1 are schemata over lexical entries. An alternative is to think of them as genuine type variables that should really be abstracted over as in System F [7], such that e.g. the translation for *and* would be as shown in (18), with the universally-quantified type indicated.

(18) $\Lambda\alpha.\Lambda\beta.\Lambda\gamma.\lambda p.\lambda q.\lambda i.\lambda a.p(i)([a]_0) \wedge q(i,[a]_0)([a]_1):$
$\forall\alpha.\forall\beta.\forall\gamma.(\alpha \to \beta \to t) \to (\alpha \times \beta \to \gamma \to t) \to \alpha \to \beta \times \gamma \to t$

What I want to stress, though, is that in either case we do not need the whole power of System F, since we only need (and want) the type variables to range over unquantified types. [5] has provided a set-theoretic model theory for this kind of polymorphism, whereas there are no set-theoretic models for the full System F [19].

$$
\begin{aligned}
\textit{and} &\mapsto \lambda p^{\alpha\to\beta\to t}.\lambda q^{\alpha\times\beta\to\gamma\to t}.\lambda i^{\alpha}.\lambda a^{\beta\times\gamma}.p(i)([a]_0) \wedge q(i,[a]_0)([a]_1) \\
\textit{if} &\mapsto \lambda p^{\alpha\to\beta\to t}.\lambda q^{\alpha\times\beta\to\gamma\to t}.\lambda i^{\alpha}.\lambda f^{\beta\to\gamma}.\forall x^{\beta}.p(i)(x) \to q(i,x)(f(x)) \\
\textit{not} &\mapsto \lambda Q^{e\to\alpha\to\beta\to t}.\lambda x^{e}.\lambda i^{\alpha}.\lambda f^{\beta\to\beta}.\forall b^{\beta}.Q(x)(i)(b) \to f(b) \neq f(b) \\
a &\mapsto \lambda P.\lambda V.\lambda i^{\beta}.\lambda a^{(e\times\alpha)\times\gamma}.P([a]_0)(i) \wedge V([[a]_0]_0)(i,[a]_0)([a]_1) \\
\textit{every} &\mapsto \lambda P.\lambda V.\lambda i^{\beta}.\lambda f^{(e\times\alpha)\to\gamma}.\forall a^{e\times\alpha}.P(a)(i) \to V([a]_0)(i,a)(f(a)) \\
\textit{who} &\mapsto \lambda V.\lambda P.\lambda a^{e\times\alpha\times\gamma}.\lambda i^{\beta}.P([a]_0,[[a]_1]_0)(i) \wedge V([a]_0)(i,([a]_0,[[a]_1]_0))([[a]_1]_1) \\
&\quad \text{where } P : e\times\alpha\to\beta\to t \text{ and } V : e\to\beta\times e\times\alpha\to\gamma\to t \\
\textit{donkey} &\mapsto \lambda a^{e\times 1}.\lambda i^{\alpha}.\mathsf{donkey}([a]_0) \\
\textit{brays} &\mapsto \lambda x^{e}.\lambda i^{\alpha}.\lambda e^{v}.\mathsf{bray}(x,e) \\
\textit{owns} &\mapsto \lambda D^{(e\to\alpha\to v\to t)\to\beta\to\gamma\to t}.\lambda x^{e}.D\big(\lambda y^{e}.\lambda a^{\alpha}.\lambda e^{v}.\mathsf{own}(x,y,e)\big) \\
\textit{regrets} &\mapsto \lambda D^{(v\to\alpha\to v\to t)\to\beta\to\gamma\to t}.\lambda x^{e}.D\big(\lambda d^{v}.\lambda a^{\alpha}.\lambda e^{v}.\mathsf{regret}(x,d,e)\big) \\
\textit{Giles} &\mapsto \lambda P^{e\to\alpha\to\beta\to t}.\lambda i^{\alpha}.\lambda a^{e\times\beta}.P([a]_0)(i)([a]_1) \wedge [a]_0 = \mathsf{giles} \\
\textit{he} &\mapsto \lambda V^{e\to\alpha\to\beta\to t}.\lambda i^{\alpha}.V(g^{\alpha\to e}(i))(i) \\
\textit{it} &\mapsto \lambda V^{\alpha\to\beta\to\gamma\to t}.\lambda i^{\beta}.V(g^{\beta\to\alpha}(i))(i) \\
&\quad \text{where } g \text{ stands for an arbitrarily-chosen free variable}
\end{aligned}
$$

Fig. 1. Some (schematic) lexical entries

Secondly, note that all the lexical entries incorporate an extra (polymorphic) argument position for input context, and that context is updated and passed on in appropriate ways. For example, the lexical entry for *and* requires that the first conjunct, and the conjunction as a whole, be dependent on input context i of (some) type α. The second conjunct is then dependent on i extended with the contribution of the first conjunct, of type β. The effect of this will be seen in the treatment of examples to be considered. N.B. for the sake of transparency, the lexical entry given for conjunction is 'leftward-looking', i.e. the first argument is the first conjunct.

Thirdly, the lexical entry for the common noun has a 'dummy' position filled by abstraction over the unit type. This is so as to achieve uniformity for modified and unmodified common nouns: both *donkey* and *donkey who brays* will be of type $e \times \alpha \rightarrow \beta \rightarrow t$, for some α and β, and hence the issue of type mismatch raised in Sect. 2.1 with respect to these examples does not arise.[6]

Fourthly and finally, the lexical entries given for the pronouns *he* and *it* contain a free variable. The idea is that this free variable is resolved in context, subject to constraints that will be discussed. This aspect of the analysis is certainly not crucial: [11] has shown how, with the appropriate syntax, apparently free variables can actually be lambda-bound and retained in interpretation until such point as they are discharged. In the interest of making as few assumptions about syntax as possible, however, I retain a free variable analysis.

2.3 Examples

Now let us consider some examples, beginning with the example of cross-sentential binding shown in (19), where *it* is most readily interpreted as roughly synonymous with *the donkey that brays.*

(19) A donkey brays. Giles owns it.

We assume that the input context for (19) contains no information, so it is $* : 1$.[7] In what follows I will use the notation $[word]^{var:=type}$, meaning the translation of *word* on the assumption that the type metavariable *var* is resolved to *type*.

The interpretation of the first sentence of (19) proceeds as follows:

$$\textit{a donkey} \mapsto [a]^{\alpha,\beta:=1;\gamma:=v}\left([donkey]^{\alpha:=1}\right)$$
$$\Rightarrow_\beta \lambda V^{e \rightarrow 1 \times e \times 1 \rightarrow v \rightarrow t}.\lambda i^1.\lambda a^{(e \times 1) \times v}.\mathsf{donkey}([[a]_0]_0) \wedge V([[a]_0]_0)(i,[a]_0)([a]_1)$$
$$\textit{a donkey brays} \mapsto [a\ donkey]\left([brays]^{\alpha:=1 \times e \times 1}\right)$$
$$\Rightarrow_\beta \lambda i^1.\lambda a^{(e \times 1) \times v}.\mathsf{donkey}([[a]_0]_0) \wedge \mathsf{bray}([[a]_0]_0,[a]_1)$$

The interpretation is not dependent on the input context: λi is a vacuous abstraction. The second sentence is then interpreted as follows:

$$\textit{owns it} \mapsto [owns]^{\alpha,\beta:=1 \times (e \times 1) \times v;\gamma:=v}\left([it]^{\alpha:=e;\beta:=1 \times (e \times 1) \times v;\gamma:=v}\right)$$
$$\Rightarrow_\beta \lambda x^e.\lambda i^{1 \times (e \times 1) \times v}.\lambda e^v.\mathsf{own}(x, g^{1 \times (e \times 1) \times v \rightarrow e}(i), e)$$
$$\textit{Giles owns it} \mapsto [Giles]^{\alpha:=1 \times (e \times 1) \times v;\beta:=v}([owns\ it])$$
$$\Rightarrow_\beta \lambda i^{1 \times (e \times 1) \times v}.\lambda a^{e \times v}.\mathsf{own}([a]_0, g^{1 \times (e \times 1) \times v \rightarrow e}(i),[a]_1) \wedge [a]_0 = \mathsf{giles}$$

This time the interpretation is dependent on the input context because of the pronoun: λi is not a vacuous abstraction. Putting the sentences together, we have:

---

[6] The same issue prompted [2] to switch from treating common nouns as type-denoting to predicate-denoting.

[7] In the rest of the paper this will be referred to as 'the null context', and will generally be assumed.

$$(19) \mapsto [\![and]\!]^{\alpha:=1;\beta:=(e\times 1)\times v;\gamma:=e\times v}([\![a\ donkey\ brays]\!])([\![Giles\ owns\ it]\!])$$
$$\begin{aligned}\Rightarrow_\beta \lambda i^1.\lambda a^{((e\times 1)\times v)\times e\times v}.(&\mathsf{donkey}([[[a]_0]_0]_0) \wedge \mathsf{bray}([[[a]_0]_0]_0, [[a]_0]_1))\\ &\wedge (\mathsf{own}([[a]_1]_0, g^{1\times(e\times 1)\times v\to e}(i,[a]_0), [[a]_1]_1)\\ &\wedge [[a]_1]_0 = \mathsf{giles})\end{aligned}$$

Now the interpretation is potentially dependent on the input context, because i is within the argument to the free variable g. However, g is also fed $[a]_0$, which means that *it* in (19) can refer back to a referent introduced in the first clause.

g is a free variable and is contextually resolved. However, we can impose constraints on what a natural resolution would be. What we want to say is that if g is a function the domain of which is a tuple (of tuples...), then a natural resolution of g is a function that selects an element of (an element of...) that tuple. This requirement can be given the recursive definition shown in (20).

(20) For any types α, β and γ:
   a. $\lambda b^\alpha.b$ is a natural resolution function (NRF).
   b. $\lambda b^{\alpha\times\beta}.[b]_0$ is an NRF.
   c. $\lambda b^{\alpha\times\beta}.[b]_1$ is an NRF.
   d. For any terms $F:\beta\to\gamma$ and $G:\alpha\to\beta$, $\lambda b^\alpha.F(G(b))$ is an NRF if F and G are NRFs.

So in particular, $\lambda b^{1\times(e\times 1)\times v}.[[[b]_1]_0]_0$ is a natural resolution function. With this resolution, (19) would be interpreted as shown in (21).

$$\begin{aligned}(21)\quad &\lambda i^1.\lambda a^{((e\times 1)\times v)\times e\times v}.(\mathsf{donkey}([[[a]_0]_0]_0) \wedge \mathsf{bray}([[[a]_0]_0]_0, [[a]_0]_1))\\ &\qquad \wedge (\mathsf{own}([[a]_1]_0, \lambda b([[[b]_1]_0]_0)(i,[a]_0), [[a]_1]_1)\\ &\qquad \wedge [[a]_1]_0 = \mathsf{giles})\\ \Rightarrow_\beta\quad &\lambda i^1.\lambda a^{((e\times 1)\times v)\times e\times v}.(\mathsf{donkey}([[[a]_0]_0]_0) \wedge \mathsf{bray}([[[a]_0]_0]_0, [[a]_0]_1))\\ &\qquad \wedge (\mathsf{own}([[a]_1]_0, [[[a]_0]_0]_0, [[a]_1]_1) \wedge [[a]_1]_0 = \mathsf{giles})\end{aligned}$$

This is the interpretation of the two-sentence discourse shown in (19). It is a relation, as in common in dynamic semantic systems, between an input and an output. It is also common in dynamic semantic systems to give a derived truth definition for this relational meaning. What that amounts to in this case is to take the input to be the null context $* : 1$ (as discussed above), and then existentially close the result;[8] as noted in Sect. 2.1, existential closure achieves the effect of the non-empty type condition from TTS. If we do that, then we derive (22) from (21).

[8] This corresponds closely to the truth definition for DRT proposed in [12], p. 149.

$$
\begin{array}{ll}
(22) & \exists a^{((e \times 1) \times v) \times e \times v}.(\mathsf{donkey}([[[a]_0]_0]_0) \wedge \mathsf{bray}([[[a]_0]_0]_0, [[a]_0]_1)) \\
& \qquad \wedge \left(\mathsf{own}([[a]_1]_0, [[[a]_0]_0]_0, [[a]_1]_1) \wedge [[a]_1]_0 = \mathsf{giles}\right) \\
\equiv & \exists x^e.\exists e^v.\exists y^e.\exists {e_1}^v.(\mathsf{donkey}(x) \wedge \mathsf{bray}(x, e)) \wedge \left(\mathsf{own}(y, x, e_1) \wedge y = \mathsf{giles}\right)
\end{array}
$$

(22) accurately represents the intended interpretation of (19).

The system provides a semantic account for why the interpretation of *it* cannot covary with donkeys in (23) or (the most natural interpretation of)[9] (24).

(23) Every donkey brays. Giles owns it.

(24) Giles does not own a donkey. It brays.

(23) is interpreted as follows:

$$
\begin{array}{l}
[\![and]\!]^{\substack{\alpha:=1;\beta:=\tau;\\ \gamma:=e \times v}} \quad \left([\![every]\!]^{\alpha,\beta:=1}\left([\![donkey]\!]^{\alpha:=1}\right)\left([\![brays]\!]^{\alpha:=1 \times e \times 1}\right)\right) \\
\qquad \left([\![Giles]\!]^{\alpha:=1 \times \tau;\beta:=v}\left([\![owns]\!]^{\alpha,\beta:=1 \times \tau;\gamma:=v}\left([\![it]\!]^{\alpha:=e;\beta:=1 \times \tau;\gamma:=v}\right)\right)\right) \\
\Rightarrow_\beta \lambda i^1.\lambda a^{\tau \times e \times v}.\forall x^{e \times 1}(\mathsf{donkey}([x]_0) \rightarrow \mathsf{bray}([x]_0, [a]_0(x))) \\
\qquad \wedge \left(\mathsf{own}([[a]_1]_0, g^{1 \times \tau \rightarrow e}(i, [a]_0), [[a]_1]_1) \wedge [[a]_0]_0 = \mathsf{giles}\right)
\end{array}
$$

Where $\tau := e \times 1 \rightarrow v$. Given the type of the free variable g, there is no way to get a bound reading for the pronoun.

The explanation in the case of (24) is essentially the same. It is (most naturally) interpreted as follows:

$$
\begin{array}{l}
[\![and]\!]^{\alpha:=1;\beta:=e \times (\tau \rightarrow \tau);\gamma:=v} \\
\quad \left([\![Giles]\!]^{\substack{\alpha:=1;\\ \beta:=\tau \rightarrow \tau}}\left([\![not]\!]^{\substack{\alpha:=1;\\ \beta:=\tau}}\left([\![own]\!]^{\substack{\alpha:=1 \times e \times 1;\\ \beta:=1;\gamma:=\tau}}\left([\![a]\!]^{\substack{\alpha,\beta:=1;\\ \gamma:=v}}\left([\![donkey]\!]^{\alpha:=1}\right)\right)\right)\right)\right) \\
\quad \left([\![it]\!]^{\alpha:=e;\beta:=1 \times e \times (\tau \rightarrow \tau);\gamma:=v}\left([\![brays]\!]^{\alpha:=1 \times e \times (\tau \rightarrow \tau)}\right)\right) \\
\Rightarrow_\beta \lambda i^1.\lambda a^{(e \times (\tau \rightarrow \tau)) \times v}.\Big(\forall b^\tau\left((\mathsf{donkey}([[b]_0]_0) \wedge \mathsf{own}([[a]_0]_0, [[b]_0]_0, [b]_1)\right) \\
\qquad\qquad \rightarrow [[a]_0]_1(b) \neq [[a]_0]_1(b)) \\
\qquad\qquad \wedge [[a]_0]_0 = \mathsf{giles}\Big) \wedge \mathsf{bray}(g^{(1 \times e \times (\tau \rightarrow \tau)) \rightarrow e}(i, [a]_0), [a]_1)
\end{array}
$$

Where $\tau := (e \times 1) \times v$. Once again, given the type of the free variable g, there is no way to get an unattested bound reading for the pronoun.

As in TTS, the treatment of negation here[10] is inspired by the equivalence, in classical and intuitionistic logic, of $\neg\phi$ and $\phi \rightarrow \bot$. A proof object of *Giles*

[9] Some speakers may allow an interpretation of (24) on which *a donkey* takes wider scope than negation. In that case, the pronoun could anaphorically refer back to the donkey.

[10] Figure 1 defines VP negation, which is derived from sentential negation in the obvious way. The VP formulation is more transparent in terms of compositional semantics, and also makes Giles available for anaphoric reference.

does not own a donkey is taken to be a pair consisting of Giles and a function mapping every state of Giles owning a donkey to an absurd (non-self-identical) object. Therefore, for there to be a proof object of *Giles does not own a donkey*, there cannot be a proof object of *Giles owns a donkey.*

Now we can look at a couple of genuine donkey sentences.

(25) Every farmer who owns a donkey feeds it.

$$\begin{aligned}
&\mapsto [every]^{\substack{\alpha:=1\times\tau;\\ \beta:=1;\gamma:=v}} \left([who]^{\substack{\alpha,\beta:=1;\\ \gamma:=\tau}} \left([owns]^{\substack{\alpha:=\sigma\times e\times 1;\\ \beta:=\sigma;\gamma:=\tau}} \left([a]^{\alpha,\beta:=\sigma}([donkey]^{\alpha:=\sigma})\right)\right)\right.\\
&\qquad\qquad\left.\left([farmer]^{\alpha:=\sigma;\beta:=1;\gamma:=v}\right)\right)\\
&\qquad\left([feeds]^{\alpha,\beta:=1\times e\times 1\times\tau;\gamma:=v}\left([it]^{\alpha:=e;\beta:=1\times e\times 1\times\tau;\gamma:=v}\right)\right)\\
&\Rightarrow_\beta \lambda i^1.\lambda f^{e\times 1\times\tau\rightarrow v}.\forall a^{e\times 1\times\tau}.(\mathsf{farmer}([a]_0)\wedge(\mathsf{donkey}([[[[a]_1]_1]_0]_0)\\
&\qquad\qquad\wedge\,\mathsf{own}([a]_0,[[[[a]_1]_1]_0]_0,[[[a]_1]_1]_1)))\\
&\qquad\qquad\rightarrow \mathsf{feed}([a]_0, g^{1\times\tau\rightarrow e}(i,a), f(a))
\end{aligned}$$

Where $\sigma := 1\times e\times 1$ and $\tau := (e\times 1)\times v$, as for the rest of the discussion of (25).

This time, the resolution function that we want is $\lambda b.[[[[[b]_1]_1]_1]_0]_0$.[11] If g is resolved in this way then we derive the interpretation shown in (26).

$$\begin{aligned}
(26)\quad & \lambda i^1.\lambda f^{(e\times 1)\times\tau\rightarrow v}.\forall a^{(e\times 1)\times\tau}.(\mathsf{farmer}([a]_0)\wedge(\mathsf{donkey}([[[[a]_1]_1]_0]_0)\\
&\qquad\wedge\,\mathsf{own}([a]_0,[[[[a]_1]_1]_0]_0,[[[a]_1]_1]_1)))\\
&\qquad\rightarrow \mathsf{feed}([a]_0, \lambda b([[[[[b]_1]_1]_1]_0]_0)(i,a), f(a))\\
\Rightarrow_\beta\quad & \lambda i^1.\lambda f^{(e\times 1)\times\tau\rightarrow v}.\forall a^{(e\times 1)\times\tau}.(\mathsf{farmer}([a]_0)\wedge(\mathsf{donkey}([[[[a]_1]_1]_0]_0)\\
&\qquad\wedge\,\mathsf{own}([a]_0,[[[[a]_1]_1]_0]_0,[[[a]_1]_1]_1)))\\
&\qquad\rightarrow \mathsf{feed}([a]_0, [[[[a]_1]_1]_0]_0, f(a))
\end{aligned}$$

If we apply (26) to the null context and then apply existential closure, we get (27).

$$\begin{aligned}
(27)\quad & \exists f^{(e\times 1)\times\tau\rightarrow v}.\forall a^{(e\times 1)\times\tau}.(\mathsf{farmer}([a]_0)\wedge(\mathsf{donkey}([[[[a]_1]_1]_0]_0)\\
&\qquad\wedge\,\mathsf{own}([a]_0,[a]_{1,1,0,0},[[[a]_1]_1]_1)))\\
&\qquad\rightarrow \mathsf{feed}([a]_0,[[[[a]_1]_1]_0]_0,f(a))\\
\equiv\quad & \forall x^e.\forall y^e.\forall e^v.(\mathsf{farmer}(x)\wedge(\mathsf{donkey}(y)\wedge\mathsf{own}(x,y,e)))\rightarrow\exists {e_1}^v.\mathsf{feed}(x,y,e_1)
\end{aligned}$$

The interpretation just derived is equivalent to that derived for (28), as shown below.

[11] As stated, $\lambda b.[[b]_1]_0$ is also a possible resolution function, which would have *it* varying with farmers rather than donkeys, which is obviously not a possible reading of (25). This reading could be ruled out by tweaking the lexical entry for *every*, but only at the cost of ruling out interpretations that we *do* want when we have an embedded clause. The mechanism for ruling out violations of ‘Principle B’ must come from somewhere else.

(28) If a farmer owns a donkey, he feeds it.

$$\begin{aligned}
&\mapsto [\mathit{if}]^{\substack{\alpha:=1;\\ \beta:=\tau;\\ \gamma:=v}} \left([a]^{\substack{\alpha,\beta:=1;\\ \gamma:=(e\times 1)\times v}} \left([\mathit{farmer}]^{\alpha:=1}\right) \right.\\
&\qquad\qquad \left. \left([\mathit{owns}]^{\substack{\alpha:=\sigma\times e\times 1;\\ \beta:=\sigma;\\ \gamma:=(e\times 1)\times v}} \left([a]^{\substack{\alpha:=1;\\ \beta:=\sigma;\\ \gamma:=v}} \left([\mathit{donkey}]^{\alpha:=\sigma}\right) \right)\right)\right)\\
&\qquad \left([\mathit{he}]^{\alpha:=1\times\tau} \left([\mathit{feeds}]^{\alpha,\beta:=1\times\tau;\gamma:=v} \left([\mathit{it}]^{\alpha:=e;\beta:=1\times\tau;\gamma:=v}\right)\right)\right)\\
&\Rightarrow_\beta \lambda i^1.\lambda f^{\tau\to v}.\forall x^\tau.(\mathsf{farmer}([[x]_0]_0) \wedge (\mathsf{donkey}([[[x]_1]_0]_0)\\
&\qquad\qquad \wedge\, \mathsf{own}([[x]_0]_0, [[[x]_1]_0]_0, [[x]_1]_1)))\\
&\qquad \to \mathsf{feed}(g^{1\times\tau\to e}(i,x), h^{1\times\tau\to e}(i,x), f(x))
\end{aligned}$$

Where $\sigma := 1 \times e \times 1$ and $\tau := (e \times 1) \times (e \times 1) \times v$, as for the rest of the discussion of (28).

The resolution functions that give us the desired outcome are $g := \lambda b.[[[b]_1]_0]_0$ and $h := \lambda b.[[[[b]_1]_1]_0]_0$. With these in place, and with (28) then applied to the null context and existentially closed, we derive the interpretation shown in (29).

(29)
$$\begin{aligned}
&\exists f^{\tau\to v}.\forall x^\tau.(\mathsf{farmer}([[x]_0]_0) \wedge (\mathsf{donkey}([[[x]_1]_0]_0)\\
&\qquad\qquad \wedge\, \mathsf{own}([[x]_0]_0, [[[x]_1]_0]_0, [[x]_1]_1)))\\
&\qquad \to \mathsf{feed}([[x]_0]_0, [[[x]_1]_0]_0, f(x))\\
\equiv\;& \forall x^e.\forall y^e.\forall e^v.(\mathsf{farmer}(x) \wedge (\mathsf{donkey}(y) \wedge \mathsf{own}(x,y,e))) \to \exists {e_1}^v.\mathsf{feed}(x,y,e_1)
\end{aligned}$$

The interpretations derived for (25) and (28) are thus equivalent, and in both cases it is the 'strong' interpretation that is derived; namely, that every farmer feeds every donkey that he owns. Weak readings will be discussed in the next section.

Finally, note how the use of eventualities as the equivalent of proof objects in this account, plus the polymorphism in the translation of *it*, allow for the most salient interpretation of (30) to be accounted for.

(30) If a farmer beats a donkey, he regrets it.

If we proceed as for (28) but with $[\mathit{regrets}]([\mathit{it}]^{\alpha:=v})$ instead of $[\mathit{feeds}]([\mathit{it}]^{\alpha:=e})$ and with appropriate resolution of the free variables introduced by the pronouns, the interpretation derived is as shown in (31) (where $\tau := (e \times 1) \times (e \times 1) \times v$).

(31)
$$\begin{aligned}
&\exists f^{\tau\to v}.\forall x^\tau.(\mathsf{farmer}([[x]_0]_0) \wedge (\mathsf{donkey}([[[x]_1]_0]_0)\\
&\qquad\qquad \wedge\, \mathsf{beat}([[x]_0]_0, [[[x]_1]_0]_0, [[x]_1]_1)))\\
&\qquad \to \mathsf{regret}([[x]_0]_0, [[x]_1]_1, f(x))\\
\equiv\;& \forall x^e.\forall y^e.\forall e^v.(\mathsf{farmer}(x) \wedge (\mathsf{donkey}(y) \wedge \mathsf{beat}(x,y,e))) \to \exists d^v.\mathsf{regret}(x,e,d)
\end{aligned}$$

3 Plurals, Generalized Quantifiers and Weak Readings

The lexical entries given in Fig. 1 do not cover plurals or determiners that resist analysis in first-order terms, like *most*. Some lexical entries illustrating the general approach to be taken in extending to these cases are shown in Fig. 2.

$$\begin{aligned}
&\textit{two} \mapsto \lambda P.\lambda V.\lambda i^{\beta}.\lambda X.\, |\lambda x^{e}.\exists a^{\alpha}.\exists d^{\gamma}.X((x,a),d)| = 2\\
&\qquad\qquad \wedge \forall b^{e\times\alpha}.\forall c^{\gamma}.X(b,c) \to \big(P(b)(i) \wedge V([b]_0)(i,b)(c)\big)
\end{aligned}$$

$$\begin{aligned}
&\textit{fewer than two} \mapsto\\
&\lambda P.\lambda V.\lambda i^{\beta}.\lambda X.\, |\lambda x^{e}.\exists a^{\alpha}.\exists d^{\gamma}.X((x,a),d)| < 2\\
&\qquad \wedge \forall b^{e\times\alpha}\big(\forall c^{\gamma}.X(b,c) \to \big(P(b)(i) \wedge V([b]_0)(i,b)(c)\big)\big)\\
&\qquad \wedge \neg\exists Y.\big(\forall m^{(e\times\alpha)\times\gamma}.X(m) \to Y(m)\\
&\qquad\qquad \wedge \neg\forall m^{(e\times\alpha)\times\gamma}.Y(m) \to X(m)\big)\\
&\qquad\qquad \wedge \forall b^{e\times\alpha}.\forall c^{\gamma}.Y(b,c) \to \big(P(b)(i) \wedge V([b]_0)(i,b)(c)\big)
\end{aligned}$$

$$\begin{aligned}
&\textit{most}_{\textit{weak}} \mapsto \lambda P.\lambda V.\lambda i^{\beta}.\lambda X.\, |\lambda x^{e}.\exists a^{\alpha}.\exists d^{\gamma}.X((x,a),d)| > \frac{|\lambda x^{e}.\exists a^{\alpha}.P(x,a)(i)|}{2}\\
&\qquad\qquad \wedge \forall b^{e\times\alpha}.\forall c^{\gamma}.X(b,c) \to \big(P(b)(i) \wedge V([b]_0)(i,b)(c)\big)
\end{aligned}$$

$$\begin{aligned}
&\textit{most}_{\textit{strong}} \mapsto \lambda P.\lambda V.\lambda i^{\beta}.\lambda X.\, |\lambda x^{e}.\exists a^{\alpha}.\exists d^{\gamma}.X((x,a),d)| > \frac{|\lambda x^{e}.\exists a^{\alpha}.P(x,a)(i)|}{2}\\
&\qquad\qquad \wedge\ \forall y^{e}\big(\forall m^{\alpha}.\forall n^{\beta}.\big(P(y,m)(n) \wedge \exists o^{\alpha}.\exists r^{\gamma}.X((y,o),r)\big)\\
&\qquad\qquad\qquad \to \exists s^{\gamma}.X((y,m),s)\big)\\
&\qquad\qquad \wedge \forall b^{e\times\alpha}.\forall c^{\gamma}.X(b,c) \to \big(P(b)(i) \wedge V([b]_0)(i,b)(c)\big)
\end{aligned}$$

where $P : e\times\alpha\to\beta\to t$, $V : e\to\beta\times e\times\alpha\to\gamma\to t$ and $X, Y : (e\times\alpha)\times\gamma\to t$

$$\textit{them} \mapsto \lambda V^{\alpha\to\beta\to\gamma\to t}.\lambda i^{\beta}.\lambda c^{\gamma}.\forall a^{\alpha}.G^{\beta\to\alpha\to t}(i) \to V(a)(i)(c)$$

where G stands for an arbitrarily-chosen free variable

Fig. 2. More (schematic) lexical entries

The general strategy for these cases will be to set up 'witness sets' in the sense of [1], or, more precisely, higher-order functions from which witness sets can be recovered. For example, the interpretation of *two donkeys VP* will be (a function from input contexts to) a higher-order function from which sets of two donkeys can be recovered. Of course, the nature of witness sets for quantifiers that aren't monotone-increasing means that the lexical entries for those will have to be more complex so as to enforce some version of what [20] calls the 'maximal participant condition'; this is evidenced by the lexical entry shown for *fewer than two*, which requires that the witness set not be a proper subset of any other witness set also in the extension of the VP.

A further point worth noting is that, throughout this section, the denotation for nouns and verbs with plural agreement is taken to be the same as that for those with singular agreement. Relatedly, only distributive readings of plurals are derived. Considering the distributive/collective distinction at the same time as everything else would take us too far afield.

Let us consider another simple example, in (32).

(32) Two donkeys bray. Giles owns them.

$$\begin{aligned}
&\textit{two donkeys bray} \mapsto \llbracket \textit{two} \rrbracket^{\alpha,\beta:=1;\gamma:=v}\left(\llbracket \textit{donkey} \rrbracket^{\alpha:=1}\right)\left(\llbracket \textit{bray} \rrbracket^{\alpha:=1}\right)\\
&\Rightarrow_\beta \lambda i^1.\lambda X^{(e\times 1)\times \bar{v}\rightarrow t}.\left|\lambda x^e.\exists a^1.\exists d^v.X((x,a),d)\right| = 2\\
&\qquad\qquad\qquad\qquad \wedge \forall b^{e\times 1}.\forall c^v.X(b,c) \rightarrow \left(\mathsf{donkey}([b]_0) \wedge \mathsf{bray}([b]_0, c)\right)
\end{aligned}$$

$$\begin{aligned}
&\textit{Giles owns them} \mapsto\\
&\llbracket \textit{Giles} \rrbracket^{\alpha:=1\times\tau;\beta:=v}\left(\llbracket \textit{owns} \rrbracket^{\alpha,\beta:=1\times\tau;\gamma:=v}\left(\llbracket \textit{them} \rrbracket^{\alpha:=e;\beta:=1\times\tau;\gamma:=v}\right)\right)\\
&\Rightarrow_\beta \lambda i^1\times\tau.\lambda a^{e\times v}.\forall y^e\left(G^{1\times\tau\rightarrow e\rightarrow t}(i)(y) \rightarrow \left(\mathsf{own}([a]_0, y, [a]_1)\right) \wedge \mathsf{giles} = [a]_0\right)\\
&\therefore (32) \mapsto \llbracket \textit{and} \rrbracket^{\alpha:=1;\beta:=\tau;\gamma:=e\times v}\left(\llbracket \textit{two donkeys bray} \rrbracket\right)\left(\llbracket \textit{Giles owns them} \rrbracket\right)\\
&\Rightarrow_\beta \lambda i^1.\lambda a^{\tau\times e\times v}.\left|\lambda x^e.\exists y^1.\exists d^v.[a]_0((x,y),d)\right| = 2\\
&\qquad\qquad \wedge \forall b^{e\times 1}\left(\forall c^v.[a]_0(b,c) \rightarrow \left(\mathsf{donkey}([b]_0) \wedge \mathsf{bray}([b]_0,c)\right)\right)\\
&\qquad\qquad \wedge \forall y^e.G^{1\times\tau\rightarrow e\rightarrow t}(i,[a]_0)(y) \rightarrow \left(\mathsf{own}([[a]_1]_0, y, [[a]_1]_1)\right.\\
&\qquad\qquad\qquad\qquad\qquad\qquad \left.\wedge [[a]_1]_0 = \mathsf{giles}\right)
\end{aligned}$$

Where $\tau := (e\times 1)\times v \rightarrow t$.

Since we are now dealing with plural (set) entities, the definition of natural resolution functions given in (20) is inadequate. We need to extend the definition so that we can extract a set of entities from a set of tuples. Clauses d. and e. in (33), an extension of (20), achieve this.

(33) For any types α, β and γ:

a. $\lambda b^\alpha.b$ is a natural resolution function (NRF).

b. $\lambda b^{\alpha\times\beta}.[b]_0$ is an NRF.

c. $\lambda b^{\alpha\times\beta}.[b]_1$ is an NRF.

d. $\lambda b^{\alpha\times\beta\rightarrow t}.\lambda Y^\alpha.\exists Z^\beta.b(Y,Z)$ is an NRF.

e. $\lambda b^{\alpha\times\beta\rightarrow t}.\lambda Y^\alpha.\exists Z^\beta.b(Z,Y)$ is an NRF.

f. For any terms $F : \beta\rightarrow\gamma$ and $G : \alpha\rightarrow\beta$, $\lambda b^\alpha.F(G(b))$ is an NRF if F and G are NRFs.

(33) means that $\lambda b^{1\times((e\times 1)\times v\rightarrow t)}.\lambda x^e.\exists n^1.\exists e^v.[b]_1((x,n),e)$ is a natural resolution function. With G in (33) instantiated to this function, the interpretation proceeds as follows.

(34) $\lambda i^1.\lambda a^{\tau \times e \times v}.\left|\lambda x^e.\exists y^1.\exists d^v.[a]_0((x,y),d)\right| = 2$
$$\wedge \forall b^{e \times 1}(\forall c^v.[a]_0(b,c) \to (\mathsf{donkey}([b]_0) \wedge \mathsf{bray}([b]_0,c)))$$
$$\wedge \forall y^e.\exists n^1(\exists e^v.[(i,[a]_0)]_1((y,n),e))$$
$$\to (\mathsf{own}([[a]_1]_0, y, [[a]_1]_1) \wedge [[a]_1]_0 = \mathsf{giles})$$
$\Rightarrow_\beta$ $\lambda i^1.\lambda a^{\tau \times e \times v}.\left|\lambda x^e.\exists y^1.\exists d^v.[a]_0((x,y),d)\right| = 2$
$$\wedge \forall b^{e \times 1}(\forall c^v.[a]_0(b,c) \to (\mathsf{donkey}([b]_0) \wedge \mathsf{bray}([b]_0,c)))$$
$$\wedge \forall y^e.\exists n^1(\exists e^v.[a]_0((y,n),e))$$
$$\to (\mathsf{own}([[a]_1]_0, y, [[a]_1]_1) \wedge [[a]_1]_0 = \mathsf{giles})$$

Where $\tau := (e \times 1) \times v \to t$.

Applied to the empty context and then existentially closed, the interpretation comes out as shown in (35).

(35) $\exists a^{((e \times 1) \times v \to t) \times e \times v}.\left|\lambda x^e.\exists y^1.\exists d^v.[a]_0((x,y),d)\right| = 2$
$$\wedge \forall b^{e \times 1}(\forall c^v.[a]_0(b,c) \to (\mathsf{donkey}([b]_0) \wedge \mathsf{bray}([b]_0,c)))$$
$$\wedge \forall y^e.\exists n^1(\exists e^v.[a]_0((y,n),e))$$
$$\to (\mathsf{own}([[a]_1]_0, y, [[a]_1]_1) \wedge [[a]_1]_0 = \mathsf{giles})$$
$\equiv$ $\exists R^{e \times v \to t}.\exists z^e.\exists e^v.|\lambda x^e.\exists d^v.R(x,d)| = 2$
$$\wedge \forall v^e(\forall c^v.R(v,c) \to (\mathsf{donkey}(v) \wedge \mathsf{bray}(v,c)))$$
$$\wedge \forall y^e.\exists b^v(R(y,b)) \to (\mathsf{own}(z,y,e) \wedge z = \mathsf{giles})$$

We are now in a position to look at a proportional donkey sentence, (36), under both strong and weak readings.

(36) Most farmers who own a donkey feed it.

With appropriate type instantiations, the weak reading of the sentence is derived as shown below.

$$\lambda i^1.\lambda X^{(e \times \tau) \times v \to t}.|\lambda x^e.\exists a^\tau.\exists d^v.X((x,a),d)| > \frac{\left|\lambda x^e.\exists a^\tau.\mathsf{farmer}(x) \wedge (\mathsf{donkey}([[[a]_1]_0]_0) \wedge \mathsf{own}(x,[[[a]_1]_0]_0,[[a]_1]_1))\right|}{2}$$
$$\wedge\ \forall b^{e \times \tau}.\forall c^v.X(b,c) \to ((\mathsf{farmer}([b]_0) \wedge (\mathsf{donkey}([[[[b]_1]_1]_0]_0) \wedge \mathsf{own}([b]_0,[[[b]_1]_1]_0]_0,[[[b]_1]_1]_1))) \wedge \mathsf{feed}([b]_0, g^{1 \times e \times \tau \to e}(i,b),c))$$

Where $\tau := 1 \times (e \times 1) \times v$, as for the rest of the discussion of (36).

In this case the resolution function that we want is $\lambda b^{1 \times e \times \tau}.[[[[b]_1]_1]_1]_0]_0$. With this resolution, followed by application to the empty context and existential closure, we end up with the interpretation shown in (37).

(37)
$$\begin{aligned}
&\exists X^{(e \times \tau) \times v \to t}. |\lambda x^e.\exists a^\tau.\exists d^v.X((x,a),d)| > \\
&\quad \frac{\left|\lambda x^e.\exists a^\tau.\mathsf{farmer}(x) \wedge \big(\mathsf{donkey}([[[a]_1]_0]_0) \wedge \mathsf{own}(x,[[[a]_1]_0]_0,[[a]_1]_1)\big)\right|}{2} \\
&\quad \wedge \forall b^{e \times \tau}.\forall c^v.X(b,c) \to \Big(\big(\mathsf{farmer}([b]_0) \wedge \big(\mathsf{donkey}([[[[b]_1]_1]_0]_0) \\
&\qquad \wedge \mathsf{own}([b]_0,[[[[b]_1]_1]_0]_0,[[[b]_1]_1]_1)\big)\big) \\
&\qquad \wedge \mathsf{feed}([b]_0,[[[[b]_1]_1]_0]_0,c)\Big) \\
\equiv\ &\exists Y^{e \times e \times v \times v \to t}. |\lambda x^e.\exists y^e.\exists d^v.\exists e^v.Y(x,y,d,e)| > \\
&\quad \frac{\left|\lambda x^e.\exists y^e.\exists e^v.\mathsf{farmer}(x) \wedge \big(\mathsf{donkey}(y) \wedge \mathsf{own}(x,y,e)\big)\right|}{2} \\
&\quad \wedge \forall x^e.\forall y^e.\forall d^v.\forall e^v.Y(x,y,d,e) \\
&\qquad \to \big(\big(\mathsf{farmer}(x) \wedge \big(\mathsf{donkey}(y) \wedge \mathsf{own}(x,y,d)\big)\big) \wedge \mathsf{feed}(x,y,e)\big)
\end{aligned}$$

Suppressing mention of eventualities for the sake of simplicity, (37) expresses the existence of a set Y of farmer-donkey pairs such that the number of farmers in Y is greater than half the number of farmers who own a donkey; and for every farmer-donkey pair in Y, the farmer owns the donkey and feeds the donkey. It does not require that every farmer in Y feed every donkey that he owns. For that, we need the lexical entry given in Fig. 2 for the strong version of *most*. With this in place, and the same resolution for the pronoun, we end up with the interpretation shown in (38).

(38)
$$\begin{aligned}
&\exists X^{(e \times \tau) \times v \to t}. |\lambda x^e.\exists a^\tau.\exists d^v.X((x,a),d)| > \\
&\quad \frac{\left|\lambda x^e.\exists a^\tau.\mathsf{farmer}(x) \wedge \big(\mathsf{donkey}([[[a]_1]_0]_0) \wedge \mathsf{own}(x,[[[a]_1]_0]_0,[[a]_1]_1)\big)\right|}{2} \\
&\quad \wedge \forall y^e \big(\forall m^\tau.\forall n^1.(\mathsf{farmer}(y) \wedge \mathsf{donkey}([[[m]_1]_0]_0) \\
&\qquad \wedge \mathsf{own}(y,[[[m]_1]_0]_0,[[m]_1]_1) \\
&\qquad \wedge \exists o^\tau.\exists r^v.X((y,o),r)) \to \exists s^v.X((y,m),s)\big) \\
&\quad \wedge \forall b^{e \times \tau}.\forall c^v.X(b,c) \to \Big(\big(\mathsf{farmer}([b]_0) \wedge \big(\mathsf{donkey}([[[[b]_1]_1]_0]_0) \\
&\qquad \wedge \mathsf{own}([b]_0,[[[[b]_1]_1]_0]_0,[[[b]_1]_1]_1)\big)\big) \\
&\qquad \wedge \mathsf{feed}([b]_0,[[[[b]_1]_1]_0]_0,c)\Big) \\
\equiv\ &\exists Y^{e \times e \times v \times v \to t}. |\lambda x^e.\exists y^e.\exists d^v.\exists e^v.Y(x,y,d,e)| > \\
&\quad \frac{\left|\lambda x^e.\exists y^e.\exists e^v.\mathsf{farmer}(x) \wedge \big(\mathsf{donkey}(y) \wedge \mathsf{own}(x,y,e)\big)\right|}{2} \\
&\quad \wedge \forall y^e \big(\forall z^e.\forall d^v.(\mathsf{farmer}(y) \wedge \mathsf{donkey}(z) \wedge \mathsf{own}(y,z,d) \\
&\qquad \wedge \exists v^e.\exists c^v.\exists r^v.Y(y,v,c,r)) \\
&\qquad \to \exists s^v.Y(y,z,d,s)\big) \\
&\quad \wedge \forall x^e.\forall y^e.\forall d^v.\forall e^v.Y(x,y,d,e) \\
&\qquad \to \big(\big(\mathsf{farmer}(x) \wedge \big(\mathsf{donkey}(y) \wedge \mathsf{own}(x,y,d)\big)\big) \wedge \mathsf{feed}(x,y,e)\big)
\end{aligned}$$

In addition to what is expressed in (37), (38) requires that for every farmer-donkey pair in Y, if the farmer owns any other donkey then that farmer-donkey pair is in Y as well. This captures the strong reading.

4 Discussion and Conclusion

I have presented a framework for capturing many anaphoric relationships which is inspired by the concepts behind TTS analyses of these phenomena, but without actually using an enriched type theory like ITT. Nevertheless, the type theory used is not exactly the simple theory of types either, as the account crucially relies on type polymorphism, either at the object-level or at the meta-level. This requirement, however, appears to be at least partly independent of the TTS/MTS distinction, since type polymorphism is also required in TTS analyses once the account is extended to include generalized quantifiers, as in [22].

On the MTS side, there is an obvious similarity between the account presented in this paper and accounts that make use of lists or stacks to keep track of discourse referents, for example [3,4,9,17]. That in itself suggests a connection between MTS and TTS approaches to anaphora, particularly between lists/stacks of discourse referents and (Martin-Löf) proof objects, that should be further explored.

References

1. Barwise, J., Cooper, R.: Generalized quantifiers and natural language. Linguist. Philos. **4**(2), 159–219 (1981)
2. Bekki, D.: Representing anaphora with dependent types. In: Asher, N., Soloviev, S. (eds.) LACL 2014. LNCS, vol. 8535, pp. 14–29. Springer, Heidelberg (2014). https://doi.org/10.1007/978-3-662-43742-1_2
3. Dekker, P.: Predicate logic with anaphora. In: Harvey, M., Samelmann, L. (eds.) Proceedings of Semantics and Linguistic Theory, vol. 4, pp. 79–95 (1994)
4. van Eijck, J.: Incremental dynamics. J. Logic Lang. Inf. **10**, 319–351 (2001)
5. Emms, M.: Polymorphic quantifiers. In: Stokhof, M., Torenvliet, L. (eds.) Proceedings of the Seventh Amsterdam Colloquium. Institute for Language, Logic and Information, Amsterdam (1989)
6. Geach, P.T.: Reference and Generality. Contemporary Philosophy. Cornell University Press, Ithaca (1962)
7. Girard, J.Y.: The system F of variable types, fifteen years later. Theor. Comput. Sci. **45**, 159–192 (1986)
8. Groenendijk, J., Stokhof, M.: Dynamic predicate logic. Linguist. Philos. **14**(1), 39–100 (1991)
9. de Groote, P.: Towards a Montagovian account of dynamics. In: Gibson, M., Howell, J. (eds.) Proceedings of Semantics and Linguistic Theory, vol. 16, pp. 1–16 (2006)
10. Heim, I.: The semantics of definite and indefinite nouns phrases. Ph.D. thesis, University of Massachussetts, Amherst (1982)
11. Jacobson, P.: Towards a variable-free semantics. Linguist. Philos. **22**(2), 117–184 (1999)

12. Kamp, H., van Genabith, J., Reyle, U.: Discourse representation theory. In: Gabbay, D.M., Guenther, F. (eds.) Handbook of Philosophical Logic, vol. 15, 2nd edn, pp. 125–394. Springer, Dordrecht (2011). https://doi.org/10.1007/978-94-007-0485-5_3
13. Kamp, H., Reyle, U.: From Discourse to Logic. Studies in Linguistics and Philosophy, vol. 42. Springer, Dordrecht (1993). https://doi.org/10.1007/978-94-017-1616-1
14. Luo, Z.: Formal semantics in modern type theories: is it model-theoretic, proof-theoretic, or both? In: Asher, N., Soloviev, S. (eds.) LACL 2014. LNCS, vol. 8535, pp. 177–188. Springer, Heidelberg (2014). https://doi.org/10.1007/978-3-662-43742-1_14
15. Martin-Löf, P.: An intuitionistic theory of types: predicative part. In: Rose, H., Shepherdson, J. (eds.) Logic Colloquium 1973. Studies in Logic and the Foundations of Mathematics, vol. 80, pp. 73–118. North-Holland, Amsterdam (1975)
16. Montague, R.: The proper treatment of quantification in ordinary English. In: Suppes, P., Moravcsik, J., Hintikka, J. (eds.) Approaches to Natural Language, pp. 221–242. D. Reidel, Dordrecht (1973)
17. Nouwen, R.: On dependent pronouns and dynamic semantics. J. Philos. Logic **36**(2), 123–154 (2007)
18. Ranta, A.: Type-Theoretical Grammar. Indices, vol. 1. Oxford University Press, Oxford (1994)
19. Reynolds, J.C.: Polymorphism is not set-theoretic. In: Kahn, G., MacQueen, D.B., Plotkin, G. (eds.) SDT 1984. LNCS, vol. 173, pp. 145–156. Springer, Heidelberg (1984). https://doi.org/10.1007/3-540-13346-1_7
20. Steedman, M.: Taking Scope. MIT Press, Cambridge (2012)
21. Sundholm, G.: Proof theory and meaning. In: Gabbay, D., Guenther, F. (eds.) Handbook of Philosophical Logic, vol. 3, pp. 471–506. D. Reidel, Dordrecht (1986)
22. Tanaka, R., Nakano, Y., Bekki, D.: Constructive generalized quantifiers revisited. In: Nakano, Y., Satoh, K., Bekki, D. (eds.) JSAI-isAI 2013. LNCS (LNAI), vol. 8417, pp. 115–124. Springer, Cham (2014). https://doi.org/10.1007/978-3-319-10061-6_8

Logical Entity Level Sentiment Analysis

Niklas Christoffer Petersen and Jørgen Villadsen(✉)

Department of Management Engineering, Department of Applied Mathematics
and Computer Science, Technical University of Denmark,
2800 Kongens Lyngby, Denmark
{niklch,jovi}@dtu.dk

Abstract. We present a formal logical approach using a combinatory categorial grammar for entity level sentiment analysis that utilizes machine learning techniques for efficient syntactical tagging and performs a deep structural analysis of the syntactical properties of texts in order to yield precise results. The method should be seen as an alternative to pure machine learning methods for sentiment analysis, which are argued to have high difficulties in capturing long distance dependencies, and can be dependent on significant amount of domain specific training data. The results show that the method yields high correctness, but further investment is needed in order to improve its robustness.

1 Introduction

The amount of unstructured textual data available through the Social Web has grown rapidly over the last years. The potential in such data are numerous, and has found applications in both commercial products and services, as well as the political and financial world cf. [6].

Sentiment analysis (or opinion mining) has enjoyed high research activity for some time now, sparked by work such as [11,14]. Traditional approaches include statistical text classifiers and keyword-based algorithms, and will usually classify sentiment on document, sentence or simply on word level. However such granularity suffers from obvious weaknesses, for instance when trying to analyze sentences with coordination of sentiments for multiple entities, e.g. (1).

The buffet was expensive, but the view is amazing. (1)

To cope with such cases the granularity of sentiment analysis has in recent work shifted towards entity level (or concept level) approaches. However this increased degree of detail introduces new challenges, especially for statistical methods, cf. [3], both due to their semantic weakness and because labeled training data are sparse at this granularity level. Statistical methods generally rely on some fixed window for feature extraction (i.e. n-grams), and can thus fail to detect long distance dependencies between an entity and opinion stated about that entity. An illustration of this is shown by the potentially unbound number of *relative clauses* allowed in English, e.g. (2), where *breakfast* is described as

A. Foret et al. (Eds.): FG 2017, LNCS 10686, pp. 54–71, 2018.
https://doi.org/10.1007/978-3-662-56343-4_4

best, however one would need to use a window size of at least 9 to detect this relation, which is arguably larger then normally considered ([11] only considers up to tri-grams).

The breakfast that was served Friday morning was the best I ever had! (2)

We present a formal logical approach for entity level sentiment analysis that utilizes machine learning techniques for efficient syntactical tagging and performs a deep structural analysis of the syntactical properties of texts in order to yield precise results.

The present paper is a substantial extension of [15]. Specifically we elaborate on the semantic annotation and the use of semantic networks for assignment of sentiment polarity. Furthermore, we extend the method with intensifiers and qualifiers, i.e. adverbs that respectively strengthen or weaken the sentiment.

After Sect. 2 on related work we present in Sect. 3 the combinatory categorial grammar and the tagging model used. Section 4 describes the adaption to sentiment analysis. The experimental results are described in Sect. 5 and further discussed in Sect. 6. Finally Sect. 7 concludes.

2 Related Work

Notably related work using formal approaches includes [20] where the authors present a method of extracting sentiment from dependency structures and also focus on capturing long distance dependencies. As dependency structures simply can be seen as binary relations on words, it is indeed a formal approach. However what seems rather surprising is that in the end they only classify on sentence-level, and thus in this process loose the entity of the dependency.

The most similar work on sentiment analysis found using a formal approach is the work [17]. The paper presents a method to detect sentiment of newspaper headlines, in fact partially using the same grammar formalism that later will be introduced and used in the present work, but, however, without the combinatorial logic approach. The paper focus on some specific problems arising with analyzing newspaper headlines, e.g. such as headline texts often do not constitute a complete sentence, etc. However the paper also present more general methods, including a method for building a highly covering map from words to polarities based on a small set of positive and negative seed words. This method has been adopted by this approach as it solves the assignment of polarity values on the lexical level quite elegantly, and is very loosely coupled to the domain. However their actual semantic analysis, which unfortunately is described somewhat shallow in the paper, seem to suffer from severe problems with respect to certain phrase structures, e.g. *dependent clauses*.

Finally it is noted, that there seem to be a strong imbalance between the *formal approaches* and *machine learning approaches*, with respect to amount of research, i.e. there exists a lot of research on sentiment analysis using machine learning compared to research embracing formal methods.

3 Materials and Methods

3.1 Combinatory Categorial Grammar

The grammar formalism used is *Combinatory Categorial Grammar* (CCG), pioneered largely by [18], and later enhanced by [1] to incorporate *modalities*. CCG adds a layer of combinatory logic onto pure Categorial Grammar, which allows an elegant and succinct formation of *higher-order* semantic expressions directly from the syntactic analysis. The set of modalities used, $\mathcal{M}$, follows [1,19], where $\mathcal{M} = \{\star, \diamond, \times, \cdot\}$. The set is partially ordered cf. the lattice (3).

(3)

A CCG lexicon, $\mathcal{L}_{\text{CCG}}$, is a mapping from a lexical unit, $w \in \Sigma^\star$, to a set of 2-tuples, each containing a lexical category and semantic expression that the unit can entail cf. (4), where Γ denotes the set of lexical and phrasal categories, and Λ denotes the set of semantic expressions.

$$\mathcal{L}_{\text{CCG}} : \Sigma^\star \rightarrow \mathcal{P}(\Gamma \times \Lambda) \tag{4}$$

A category is either *primitive* or *compound*. The set of primitive categories, $\Gamma_{\text{prim}} \subset \Gamma$, is language dependent and, for the English language, it consists of S (sentence), NP (noun phrase), N (noun) and PP (prepositional phrase). Compound categories are recursively defined by the infix operators $/_\iota$ (forward slash) and $\backslash_\iota$ (backward slash), i.e. if α and β are members of Γ, then so are $\alpha/_\iota\beta$ and $\alpha\backslash_\iota\beta$. The modality of the operator, $\iota \in \mathcal{M}$, can restrict the application of inference rules during deduction in order to ensure the soundness of the system. The partial ordering allows the most restrictive categories to also be included in the less restrictive, e.g. any rule that assumes $\alpha/_\diamond\beta$ will also be valid for $\alpha/.\beta$. Since $\cdot$ permits any rule it is convenient to simply write $/$ and $\backslash$ instead of respectively $/.$ and $\backslash.$, i.e. the dot is omitted from these operators.

3.2 Combinatory Rules

CCGs can be seen as a logical deductive proof system where the axioms are members of $\Gamma \times \Lambda$. A text $T \in \Sigma^\star$ is accepted as a sentence in the language, if there, for some tagging of T, exists a deductive proof for S (sentence). A tagging of a text is, for each lexical unit $w \in \Sigma^\star$ in the text, simply the selection of one of the pairs yielded by $\mathcal{L}_{\text{CCG}}(w)$. While this seems simple, it constitutes the major computational challenge of this approach. E.g. given some ordered set of lexical units, which constitutes the text T to analyse, the number of possible combinations of taggings might be very large.

Once an appropriate selection is made rewrite is done using a language independent set of *combinatory rules*:

$$
\begin{array}{rcll}
X/_\star Y : f \quad Y : a & \Rightarrow & X : f\,a & (>) \\
Y : a \quad X\backslash_\star Y : f & \Rightarrow & X : f\,a & (<) \\
X/_\diamond Y : f \quad Y/_\diamond Z : g & \Rightarrow & X/_\diamond Z : \lambda a.f(g\,a) & (>_\mathbf{B}) \\
Y\backslash_\diamond Z : g \quad X\backslash_\diamond Y : f & \Rightarrow & X\backslash_\diamond Z : \lambda a.f(g\,a) & (<_\mathbf{B}) \\
X : a & \Rightarrow & T/(T\backslash X) : \lambda f.fa & (>_\mathbf{T}) \\
X : a & \Rightarrow & T\backslash(T/X) : \lambda f.fa & (<_\mathbf{T}) \\
X/_\times Y : f \quad Y\backslash_\times Z : g & \Rightarrow & X\backslash_\times Z : \lambda a.f(g\,a) & (>_{\mathbf{B}_\times}) \\
Y/_\times Z : g \quad X\backslash_\times Y : f & \Rightarrow & X/_\times Z : \lambda a.f(g\,a) & (<_{\mathbf{B}_\times})
\end{array}
$$

Where X, Y, Z and T are variables ranging over categories (i.e. Γ), and f, a and g are variables over semantic expressions (i.e. Λ).

With only functional application, ($>$) and ($<$), the system is capable of capturing any context-free language cf. [18]. Figure 1 shows the deduction of S from the simple declarative sentence "the hotel had an exceptional service" (semantics are omitted).

the $NP/_\diamond N$ **hotel** N ⟶(>) NP

had $(S\backslash NP)/NP$

an $NP/_\diamond N$ **exceptional** N/N **service** N

N/N N ⟶(>) N

$NP/_\diamond N$ N ⟶(>) NP

$(S\backslash NP)/NP$ NP ⟶(>) $S\backslash NP$

NP $S\backslash NP$ ⟶(<) S

Fig. 1. Deduction of simple declarative sentence.

The functional composition, ($>_\mathbf{B}$) and ($<_\mathbf{B}$), is often used in connection with type-raising ($>_\mathbf{T}$) and ($<_\mathbf{T}$), for instance to allow relative clauses, coordination, while crossed functional composition, ($>_{\mathbf{B}_\times}$) and ($<_{\mathbf{B}_\times}$), are needed for more exotic linguistic phenomenons such as *heavy noun phrase shifting*.

3.3 Maximum Entropy Tagging

There exists some wide covering CCG lexicons, most notable *CCGbank*, compiled by [10] by techniques presented by [9]. It is essentially a translation of almost the entire Penn Treebank [12], which contains over 4.5 million lexical units, and where each sentence structure has been analyzed in full and annotated. The result is a highly covering lexicon, with some entries having assigned over 100 different lexical categories. Clearly such lexicons only constitutes half of the previous defined $\mathcal{L}_{\text{CCG}}$ map, i.e. only the lexical categories, Γ. The problem of obtaining semantic expressions, is addressed later.

To ensure the presented method works with a large vocabulary and a wide range of sentence structures, and thus the variety of opinion texts harvested from social networks, an efficient syntactical CCG-tagging is required. This is substantiated by the fact that [10] calculates that the expected number of lexical categories per token is 19.2 for the CCGBank. This mean that an exhaustive search of even a short sentence (seven tokens) is expected to consider over 960 million (19.2^7) possible taggings.

Machine learning is used to handle this otherwise exponentially bounded search, specifically [4] presents a method based on a *maximum entropy model* that estimates the probability that a token is to be assigned a particular category, given the *features* of the local context, e.g. the POS-tag of the current and adjacent lexical units, and the CCG-category of lexical units left to the current.

This is used to select a subset of possible categories for a lexical unit, by selecting categories with a probability within a factor of the category with highest probability. In some cases this of cause will prune the correct tagging needed to deduct S, but [4] shows that the average number of lexical categories per lexical unit can be reduced to 3.8 while the method still recognize 98.4% of unseen data.

A complete parser is presented in [5]. It utilizes this tagging model and a series of (log-linear) models to speed-up the actual deduction once the tagging model has assigned a set of categories to each token.

Finally, since the tagging models are based on trained data, which also can contain minor grammatical errors and misspellings, it is still able to assign categories to lexical entries even though they might be incorrect spelled or of wrong form, which it not very uncommon when harvesting data from social networks, user reviews, etc.

4 Theory and Calculation

4.1 Definition of a Sentiment Analysis

The sentiment polarity model used in this paper is continuous, and can thus be seen as a weighted classification. Thereby the polarity is a value in some predefined interval, $[-\omega; \omega]$. An opinion with value close to $-\omega$ is considered highly negative, whereas a value close to ω is considered highly positive. Opinions with values close to zero are considered almost neutral. This model allows the overall process of the sentiment analysis presented in this paper to be given by Definition 1.

Definition 1. *A sentiment analysis $\mathcal{A}$ is a computation on a text $T \in \Sigma^\star$ with respect to a* subject of interest $s \in \mathbb{E}$, *where $\Sigma^\star$ denotes the set of all texts, and $\mathbb{E}$ is the set of all entities. The result is a normalized score as shown in (5). The yielded score should reflect the* polarity *of the given subject of interest in the text, i.e. whether the overall opinion is positive, negative, or neutral.*

$$\mathcal{A} : \Sigma^\star \to \mathbb{E} \to [-\omega; \omega] \tag{5}$$

4.2 Combinatory Categorial Grammar for Sentiment Analysis

In order to apply the CCG formalism to the area of sentiment analysis the expressive power of the semantics needs to be adapted to this task. The set of semantic expressions, Λ, is defined as a superset of simply typed λ-expressions cf. Definition 2.

Definition 2. *Besides variables, functional abstraction and functional application, which follows from simply typed λ-expressions cf. [2], the following structures are available:*

- *A n-ary* functor *($n \geq 0$) with name f from an infinite set of functor names, polarity $j \in [-\omega; \omega]$, and* impact argument *k ($0 \leq k \leq n$).*
- *A* sequence *of n semantic expressions of the* same *type.*
- *The* change of impact argument.
- *The* change *of an expression's polarity.*
- *The* scale *of an expression's polarity. The magnitude of which an expression's polarity may scale is given by $[-\psi; \psi]$.*

Formally this can be stated:

$$
\begin{array}{rcll}
x : \tau \in \mathcal{V} & \Rightarrow & x : \tau \in \Lambda & \text{(Variable)} \\
x : \tau_\alpha \in \mathcal{V},\ e : \tau_\beta \in \Lambda & \Rightarrow & \lambda x.e : \tau_\alpha \rightarrow \tau_\beta \in \Lambda & \text{(Abstraction)} \\
e_1 : \tau_\alpha \rightarrow \tau_\beta \in \Lambda,\ e_2 : \tau_\alpha \in \Lambda & \Rightarrow & (e_1 e_2) : \tau_\beta \in \Lambda & \text{(Application)} \\
e_1, \ldots, e_n \in \Lambda, 0 \leq k \leq n,\ j \in [-\omega; \omega] & \Rightarrow & f_j^k(e_1, \ldots, e_n) \in \Lambda & \text{(Functor)} \\
e_1 : \tau, \ldots, e_n : \tau \in \Lambda & \Rightarrow & \langle e_1, \ldots, e_n \rangle : \tau \in \Lambda & \text{(Sequence)} \\
e : \tau \in \Lambda, 0 \leq k' & \Rightarrow & e^{\rightsquigarrow k'} : \tau & \text{(Impact change)} \\
e : \tau \in \Lambda,\ j \in [-\omega; \omega] & \Rightarrow & e_{\circ j} : \tau \in \Lambda & \text{(Change)} \\
e : \tau \in \Lambda,\ j \in [-\psi; \psi] & \Rightarrow & e_{\bullet j} : \tau \in \Lambda & \text{(Scale)}
\end{array}
$$

The semantics includes normal α-conversion and β-, η-reduction as shown in the semantic rewrite rules for the semantic expressions given by Definition 3. More interesting are the rules that actually allow the binding of polarities to the phrase structures. The *change of a functor* itself is given by the rule (FC1), which applies to functors with, impact argument, $k = 0$. For any other value of k the functor acts like a non-capturing enclosure that passes on any change to its k'th argument as follows from (FC2). The *change of a sequence* of expressions is simply the change of each element in the sequence cf. (SC). Finally, it is allowed to *push change* inside an abstraction as shown in (PC), simply to ensure the applicability of the β-reduction rule. Completely analogous rules are provided for the scaling as shown in respectively (FS1), (FS2), (SS) and (PS). Finally the *change of impact* allows change of a functors impact argument cf. (IC). Notice that these *change*, *scale*, *push* and *impact change* rules are type preserving, and for readability type annotation is omitted from these rules.

Definition 3. *The rewrite rules of the semantic expressions are given by the following, where* $e_1[x \mapsto e_2]$ *denotes the* safe *substitution of* x *with* e_2 *in* e_1*, and* $FV(e)$ *denotes the set of free variables in* e*. For details, see for instance, [2].*

$$
\begin{array}{rcll}
(\lambda x.e) : \tau & \Rightarrow & (\lambda y.e[x \mapsto y]) : \tau & y \notin FV(e) \quad (\alpha) \\
((\lambda x.e_1) : \tau_\alpha \to \tau_\beta)\,(e_2 : \tau_\alpha) & \Rightarrow & e_1[x \mapsto e_2] : \tau_\beta & (\beta) \\
(\lambda x.(e\,x)) : \tau & \Rightarrow & e : \tau & x \notin FV(e) \quad (\eta) \\
f_j^0(e_1, \ldots, e_n)_{\circ j'} & \Rightarrow & f^0_{j \widehat{+} j'}(e_1, \ldots, e_n) & \text{(FC1)} \\
f_j^k(e_1, \ldots, e_n)_{\circ j'} & \Rightarrow & f_j^k(e_1, \ldots, e_{k \circ j'}, \ldots, e_n) & \text{(FC2)} \\
\langle e_1, \ldots, e_n \rangle_{\circ j'} & \Rightarrow & \langle e_{1 \circ j'}, \ldots, e_{n \circ j'} \rangle & \text{(SC)} \\
(\lambda x.e)_{\circ j'} & \Rightarrow & \lambda x.(e_{\circ j'}) & \text{(PC)}
\end{array}
$$

$$
\begin{array}{rcll}
f_j^0(e_1, \ldots, e_n)_{\bullet j'} & \Rightarrow & f^0_{j \widehat{\cdot} j'}(e_1, \ldots, e_n) & \text{(FS1)} \\
f_j^k(e_1, \ldots, e_n)_{\bullet j'} & \Rightarrow & f_j^k(e_1, \ldots, e_{k \bullet j'}, \ldots, e_n) & \text{(FS2)} \\
\langle e_1, \ldots, e_n \rangle_{\bullet j'} & \Rightarrow & \langle e_{1 \bullet j'}, \ldots, e_{n \bullet j'} \rangle & \text{(SS)} \\
(\lambda x.e)_{\bullet j'} & \Rightarrow & \lambda x.(e_{\bullet j'}) & \text{(PS)}
\end{array}
$$

$$
f_j^k(e_1, \ldots, e_n)^{\leadsto k'} \Rightarrow f_j^{k'}(e_1, \ldots, e_n) \qquad \text{(IC)}
$$

It is assumed that the addition and multiplication operator, respectively $\widehat{+}$ and $\widehat{\cdot}$, always yields a result within $[-\omega; \omega]$ cf. Definition 4.

Definition 4. *The operators* $\widehat{+}$ *and* $\widehat{\cdot}$ *are defined cf. (6) and (7) such that they always yield a result in the range* $[-\omega; \omega]$*, even if the pure addition and multiplication might not be in this range.*

$$
j \widehat{+} j' = \begin{cases} -\omega & \text{if } j + j' < -\omega \\ \omega & \text{if } j + j' > \omega \\ j + j' & \text{otherwise} \end{cases} \tag{6}
$$

$$
j \widehat{\cdot} j' = \begin{cases} -\omega & \text{if } j \cdot j' < -\omega \\ \omega & \text{if } j \cdot j' > \omega \\ j \cdot j' & \text{otherwise} \end{cases} \tag{7}
$$

The presented definition of semantic expressions allows the binding between expressed sentiment and entities in the text to be analyzed, given that each lexicon entry have associated the proper expression.

Example 1 shows how to apply this for a simple declarative sentence, while Example 2 considers an example with long distance dependencies.

Example 1. This example considers simple declarative sentence (8) including semantics.

$$\textit{the hotel had an exceptional service} \tag{8}$$

The lexicon used in the example is provided in Table 1. Notice that nouns, verbs, etc. are reduced to their lemma for functor naming.

Table 1. Lexicon used in Example 1.

Token	Category	Type
the	NP_{nb}/N	$: \lambda x.x$
hotel	N	$: \text{hotel}_0$
had	$(S_{dcl}\backslash NP)/NP$	$: \lambda x.\lambda y.\text{have}_0^0(x, y)$
an	NP_{nb}/N	$: \lambda x.x$
exceptional	N/N	$: \lambda x.(x_{\circ 40})$
service	N	$: \text{service}_0$

Figure 2 shows the entity "service" is modified by the adjective "exceptional" which is immediately to the left of the entity. The semantic expression associated to "service" is simply the zero-argument functor, initial with a neutral sentiment value. The adjective has the "changed identity function" as expression with a change value of 40. Upon application of combinatorial rules, semantic expressions are reduced based on the rewrite rules given in Definition 3.

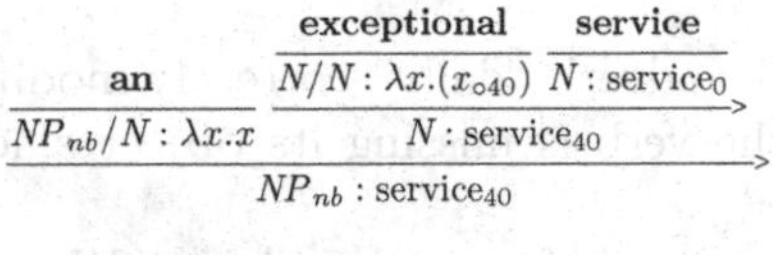

$$\dfrac{\dfrac{\textbf{an}}{NP_{nb}/N : \lambda x.x} \quad \dfrac{\dfrac{\textbf{exceptional}}{N/N : \lambda x.(x_{\circ 40})} \quad \dfrac{\textbf{service}}{N : \text{service}_0}}{N : \text{service}_{40}}>}{NP_{nb} : \text{service}_{40}}>$$

Fig. 2. Deduction of simple noun phrase sentence with semantics.

The conclusion of the deduction proof is a sentence with a semantic expression preserving most of the surface structure, and includes the bounded sentiment values on the functors cf. Fig. 3.

Example 2. This example considers the sentence (9) including semantics, and demonstrates variations of all combinator rules introduced.

$$\textit{the breakfast that the restaurant served daily was excellent} \tag{9}$$

The lexicon used in the example is provided in Table 2. Most interesting is the correct binding between "breakfast" and "excellent", even though these are far from each other in the surface structure of the sentence.

the hotel ... ; **had** ; **an exceptional service** ...

$$NP_{nb} : \mathrm{hotel}_0 \qquad (S_{dcl}\backslash NP)/NP : \lambda x.\lambda y.\mathrm{have}_0^0(x, y) \qquad NP_{nb} : \mathrm{service}_{40}$$

$$\xrightarrow{>} \; S_{dcl}\backslash NP : \lambda y.\mathrm{have}_0^0(\mathrm{service}_{40}, y)$$

$$\xrightarrow{<} \; S_{dcl} : \mathrm{have}_0^0(\mathrm{service}_{40}, \mathrm{hotel}_0)$$

Fig. 3. Deduction of simple declarative sentence with semantics.

Table 2. Lexicon used in Example 2.

Token	Category	Type
the	NP_{nb}/N	$: \lambda x.x$
breakfast	N	$: \mathrm{breakfast}_0$
that	$(N\backslash_{\diamond}N)/(S_{dcl}/_{\diamond}NP)$	$: \lambda x.\lambda y.((x\ y)^{\rightsquigarrow 1})$
restaurant	N	$: \mathrm{restaurant}_0$
served	$(S_{dcl}\backslash NP)/NP$	$: \lambda x.\lambda y.\mathrm{serve}_0^0(x, y)$
daily	$(S_X\backslash NP)\backslash(S_X\backslash NP)$	$: \lambda x.(x_{\circ 5})$
was	$(S_{dcl}\backslash NP)/(S_{adj}\backslash NP)$	$: \lambda x.x$
excellent	$S_{dcl}\backslash NP$	$: \lambda x.(x_{\circ 25})$

the restaurant ... ; **served** ; **daily**

$$NP_{nb} : \mathrm{restaurant}_0 \qquad (S_{dcl}\backslash NP)/NP : \lambda x.\lambda y.\mathrm{serve}_0^0(x, y) \qquad (S_X\backslash NP)\backslash(S_X\backslash NP) : \lambda x.(x_{\circ 5})$$

$$\xrightarrow{>T} \; S_X/(S_X\backslash NP) : \lambda f.(f\ \mathrm{restaurant}_0) \qquad \xrightarrow{<B_\times} \; (S_{dcl}\backslash NP)/NP : \lambda x.\lambda y.\mathrm{serve}_5^0(x, y)$$

$$\xrightarrow{>B} \; S_{dcl}/NP : \lambda x.\mathrm{serve}_5^0(x, \mathrm{restaurant}_0)$$

Fig. 4. Deduction of dependent clause.

Figure 4 shows how the adverb "daily" correctly modifies the transitive verb "served", even though the verb is missing its object since it participates in a relative clause.

Figure 5 shows the details of the relative clause. When the relative pronoun binds the dependent clause to the main clause, it "closes" it for further modification by changing the impact argument of the functor inflicted by the verb of the dependent clause, so that further modification will impact the subject of the main clause.

the ; **breakfast** ; **that** ; **the restaurant served daily** ...

$$NP_{nb}/N : \lambda x.x \qquad N : \mathrm{breakfast}_0 \qquad (N\backslash_{\diamond}N)/(S_{dcl}/_{\diamond}NP) : \lambda x.\lambda y.((x\ y)^{\rightsquigarrow 1}) \qquad S_{dcl}/NP : \lambda x.\mathrm{serve}_5^0(x, \mathrm{restaurant}_0)$$

$$\xrightarrow{>} \; N\backslash_{\diamond}N : \lambda y.\mathrm{serve}_5^1(y, \mathrm{restaurant}_0)$$

$$\xrightarrow{<} \; N : \mathrm{serve}_5^1(\mathrm{breakfast}_0, \mathrm{restaurant}_0)$$

$$\xrightarrow{<} \; NP : \mathrm{serve}_5^1(\mathrm{breakfast}_0, \mathrm{restaurant}_0)$$

Fig. 5. Binding of relative clause and noun phrase.

Finally the the long distance binding can be established as shown in Fig. 6.

$$\dfrac{\dfrac{\textbf{the breakfast that the restaurant served daily}}{\dots}}{NP : \text{serve}_5^1(\text{breakfast}_0, \text{restaurant}_0)} \quad \dfrac{\dfrac{\textbf{was}}{(S_{dcl}\backslash NP)/(S_{adj}\backslash NP) : \lambda x.x} \quad \dfrac{\textbf{excellent}}{S_{adj}\backslash NP : \lambda x.(x_{\circ 25})}}{S_{dcl}\backslash NP : \lambda x.(x_{\circ 25})}>$$
$$\overline{S_{dcl} : \text{serve}_5^1(\text{breakfast}_{25}, \text{restaurant}_0)}<$$

Fig. 6. Sentiment of sentence with long distance dependencies.

4.3 Lexicon Annotation

There is however one essential component missing from the lexicon, namely the semantic expressions. However due to the *Principle of Categorial Type Transparency* it is known exactly *what* the types of the semantic expressions should be. There are currently a total of 429 different tags in the *maximum entropy tagging model*, thus trying to handle each of these cases individually is certainly not very robust for changes in the lexical categories. The solution is to handle some cases that need special treatment, and then use a generic annotation algorithm for all other cases. Both the generic and the special case algorithms will be a transformation $(\mathcal{T}, \Sigma^\star) \rightarrow \Lambda$, where the first argument is the type, $\tau \in \mathcal{T}$, to construct, and the second argument is the lemma, $\ell \in \Sigma^\star$, of the lexicon entry to annotate. Since the special case algorithms will fallback to the generic approach, in case preconditions for the case are not met, it is convenient to start with the generic algorithm, $\mathcal{U}_{\text{GEN}}$, which is given by Definition 5.

Definition 5. *The* generic semantic annotation algorithm, $\mathcal{U}_{\text{GEN}}$ *(10), for a type τ and lemma ℓ is defined by the auxiliary function $\mathcal{U}'_{\text{GEN}}$, which takes two additional arguments, namely an infinite set of variables $\mathcal{V}$ cf. Definition 2, and an ordered set of sub-expressions (denoted A), which initially is empty.*

$$\mathcal{U}_{\text{GEN}}(\tau, \ell) = \mathcal{U}'_{\text{GEN}}(\tau, \ell, \mathcal{V}, \emptyset) \tag{10}$$

If τ is primitive, i.e. $\tau \in \mathcal{T}_{\text{prim}}$, then the generic algorithm simply return a functor with name ℓ, polarity and impact argument both set to 0, and the ordered set A as arguments. Otherwise there must exist unique values for $\tau_\alpha, \tau_\beta \in \mathcal{T}$, such that $\tau_\alpha \rightarrow \tau_\beta = \tau$, and in this case the algorithm return an abstraction of τ_α on variable $v \in V$, and recursively generates an expression for τ_β.

$$\mathcal{U}'_{\text{GEN}}(\tau, \ell, V, A) = \begin{cases} \ell_0^0(A) : \tau & \textit{if } \tau \in \mathcal{T}_{\text{prim}} \\ \lambda v.\mathcal{U}'_{\text{GEN}}(\tau_\beta, \ell, V \setminus \{v\}, A') : \tau & \textit{otherwise, where :} \end{cases}$$

$$v \in V$$

$$\tau_\alpha \rightarrow \tau_\beta = \tau$$

$$A' = \begin{cases} A[e : \tau_\alpha \rightarrow \tau_\gamma \mapsto ev : \tau_\gamma] & \textit{if } e' : \tau_\alpha \rightarrow \tau_\gamma \in A \\ A[e : \tau_\gamma \mapsto ve : \tau_\delta] & \textit{if } \tau_\gamma \rightarrow \tau_\delta = \tau_\alpha \wedge e' : \tau_\gamma \in A \\ A \cup \{v : \tau\} & \textit{otherwise} \end{cases}$$

The recursive call also removes the abstracted variable v from the set of variables, thus avoiding recursive abstractions to use it. The ordered set of sub-expressions, A, is modified cf. A′, where the notation $A[e_1 : \tau_1 \mapsto e_2 : \tau_2]$ *is the substitution of all elements in A of type* τ_1 *with* $e_2 : \tau_2$. *Note that* e_1 *and* τ_1 *might be used to determine the new value and type of the substituted elements. Since the two conditions on A′ are not mutual exclusive, if both apply the first case will be selected. The value of A′ can be explained in an informal, but possibly easier to understand, manner:*

- *If there is a least one function in A, that takes an argument of type* τ_α, *then apply v (which is known to by of type* τ_α*) to all such functions in A.*
- *If the type of v itself is a function (i.e.* $\tau_\gamma \to \tau_\delta = \tau_\alpha$*), and A contains at least one element that can be used as argument, then substitute all such arguments in A by applying them to v.*
- *Otherwise, simply append v to A.*

Clearly the generic algorithm does not provide much use with respect to extracting the sentiment of entities in the text, i.e. it only provide some safe structures that are guaranteed to have the correct type. The more interesting annotation is actually handled by the special case algorithms. How this is done is determined by a combination of the POS-tag and the category of the entry. Most of these treatments are very simple, with the handling of adjectives and adverbs being the most interesting:

- *Determiners* with simple category, i.e. NP/N, are simply mapped to the identity function, $\lambda x.x$. While determiners have high focus in other NLP tasks, such as determine if a sentence is valid, the importance does not seem significant in sentiment analysis, e.g. whether an opinion is stated about *an entity* or *the entity* does not change the overall polarity of the opinion bound to that entity.
- *Nouns* are in general just handled by the generic algorithm, however in some cases of multi-word nouns, the sub-lexical entities may be tagged with the category N/N. In these cases the partial noun is annotated with a list structure, that eventually will capture the entire noun, i.e. $\lambda x.\langle \mathcal{U}_{\text{GEN}}(\tau_{\text{N}}, \ell), x\rangle$, where ℓ is the lemma of the entity to annotate.
- *Verbs* are just as nouns in general handled by the generic algorithm, however *linking verbs* is a special case, since they relate the subject (i.e. an entity) with one or more *predicative adjectives*. Linking verbs have the category $(S_{\text{dcl}}\backslash NP)/(S_{\text{adj}}\backslash NP)$, and since the linked adjectives directly describes the subject of the phrase such verbs are simply annotated with the identity function, $\lambda x.x$.
- *Adjectives* can have a series of different categories depending on how they participate in the sentence, however most of them have the type $\tau_\alpha \to \tau_\beta$, where $\tau_\alpha, \tau_\beta \in \mathcal{T}_{\text{prim}}$. These are annotated with the *change* of the argument, i.e. $\lambda x.x_{\circ j}$, where j is a value determined based on the lemma of the adjective. Notice that this assumes implicit type conversion of the parameter from τ_α to τ_β, however since these are both primitive, this is a sane type cast. Details on how the value j is calculated are given in the next section.

- *Adverbs* are annotated in a fashion closely related to that of adjectives. However the result might either by a *change* or a *scale*, a choice determined by the lemma: normally adverbs are annotated by the change in the same manner as adjectives, however *intensifiers* and *qualifiers*, i.e. adverbs that respectively strengthens or weakens the meaning, are scaled. The next section gives further details on how this choice is made. Finally special care are taken about negating adverbs, i.e. "not", which are scaled with a value $j = -1$.
- *Prepositions* and *relative pronouns* need to change the impact argument of captured partial sentences, i.e. *preposition phrases* and *relative clauses*, such that further modification should bind to the subject of the entire phrase as were illustrated by Example 2.
- *Conjunctions* are annotated by an algorithm closely similar to $\mathcal{U}_{\text{GEN}}$, however instead of yielding a functor of arguments, the algorithm yields a list structure. This allow any modification to bind on each of the conjugated sub-phrases.

4.4 Assignment of Sentiment Polarity

An understanding of the domain of the review is needed in order to reason about the polarity of the entities present in texts to analyse. For this purpose the concept of *semantic networks* was used. Concretely the semantic network WordNet, originally presented by [13], and later presented in depth by [7]. WordNet contains a variety of relations, however for the purpose of calculating sentiment polarity values, only the following were considered interesting:

- The *similar*-relation, r_{similar}, links *closely similar* semantic concepts, i.e. concepts having almost the synonymies mensing in most contexts. The relation is present for most concepts entailed by adjectives.
- The *see also*-relation, $r_{\text{see-also}}$, links *coarsely similar* semantic concepts, i.e. concepts having a clear different meaning, but may be interpreted interchangeably for some contexts.
- The *pertainym*-relation, $r_{\text{pertainym}}$, links the adjectives from which an adverb was derived, e.g. *extreme* is the pertainym of *extremely*.

An approach similar to the one presented by [17] was used to calculate an assignment of sentiment polarity values for adjectives and adverbs: Positive and negative *seed concepts* are identified for the domain of the analysis, respectively S_{pos} and S_{neg}, e.g. as shown in (11) and (12).

$$S_{\text{pos}} = \{\text{clean}, \text{quiet}, \text{friendly}, \text{cheap}\} \tag{11}$$

$$S_{\text{neg}} = \{\text{dirty}, \text{noisy}, \text{unfriendly}, \text{expensive}\} \tag{12}$$

The calculation of the polarity *change* and/or *scale* for some *lemma*, present in the texts to analyze, is then based on the distances between the concepts yielded by the lemma and the seed concepts. To solve semantic ambiguity a rational assumption was taken that concepts stated in the texts presumably are to be interpreted within the domain given by the seed concepts, S_{pos} and S_{neg}. Thus concepts that are strongly related to one or more seed concepts should be preferred over weaklier related concepts. The solution is to select the n *closest* relations, thus reasoning greedily positively, respectively greedily negatively.

The approach for calculating sentiment polarity values for *intensifying* or *qualifying* adverbs modify the meaning of a verb, adjective, or another adverb, some special treatment are presented for this. Analog to the positive and negative concepts, sets of respectively intensifying and qualifying seed adjectives are stated, e.g. (13) and (14). Also notice that, unlike S_{pos} and S_{neg}, these sets does not rely on the domain, as they only strengthens or weakens domain specific polarities.

$$S_{\text{intensify}} = \{\text{extreme}, \text{much}, \text{more}\} \tag{13}$$

$$S_{\text{qualify}} = \{\text{moderate}, \text{little}, \text{less}\} \tag{14}$$

The distances are normalized such that the change value for some adverb or adjective with lemma, ℓ, will always be in $[-\omega; \omega]$; and for intensifying or qualifying adverbs with lemma, ℓ, the scale will always be in $\left[\frac{1}{2}; 2\right]$.

5 Results

5.1 Test Data

The test data set chosen for evaluation of the method was the *Opinosis data set* [8]. The data set consists of approximately 7000 texts from consumer reviews on a number of different topics. The topics are ranging over different product and services, from consumer electronics (e.g. GPS navigation, music players, etc.) to hotels and restaurants. They are harvested from several online resellers and service providers, including among others *Amazon* (http://www.amazon.com/) and *TripAdvisor* (http://www.tripadvisor.com/).

Since the data set is unlabeled it was chosen to label a small subset of it in order to measure the robustness and the correctness of the presented method (see Appendix). To avoid biases toward how the proof of concept system analyzes text the labeling was performed independently by two individuals who had no knowledge of how the presented solution processes texts.

As the example texts might have hinted, the subset chosen was from the set of hotel and restaurant reviews. The *subject of interest* chosen for the analysis were *hotel rooms*, and the subset was thus randomly sampled from texts with high probability of containing this entity (i.e. containing any morphological form of the noun "room").

The individuals were given a subset of 35 review texts, and should mark each text as either positive, negative or unknown *with respect to the given subject of interest.* Out of the 35 review text the two subject's positive/negative labeling agreed on 34 of them, while unknowns and disagreements were discarded. Thus the inter-human *concordance* for the test data set was 97.1%, which is very high, and would arguably drop if just a few more individuals were used for label annotation.

5.2 Evaluation Results

An entity sentiment value was considered to *agree* with the human labeling, if it had the correct sign (i.e. positive sentiment values agreed with positive labels, and negative values with negative labels). The baseline presented here is a sentence-level baseline, calculated by using the Naive Bayes Classifier available in the Natural Language Toolkit (NLTK) for Python.

The *precision* and *recall* results for both the baseline and the presented method are shown in Table 3. As seen the recall is somewhat low for the proof of concept system, which is addressed in the next section, while it is argued that precision of the system is indeed acceptable, since even humans will not reach a 100% agreement.

Table 3. Precision and recall results for proof of concept system.

	Baseline	The presented method
Precision	71.5%	92.3%
Recall	44.1%	35.3%

6 Discussion

The presented method for *entity level* sentiment analysis using deep sentence structure analysis has shown acceptable correction results, but inadequate robustness.

The biggest issue was found to be the lack of correct syntactic tagging models. It is argued that models following a closer probability distribution of review texts than the one used would have improved the robustness of the system significantly. One might think, that if syntactic labeled target data are needed, then the presented logical method really suffers the same issue as machine learning approaches, i.e. *domain dependence.* However it is argued that exactly because the models needed are of *syntactic level*, and not of *sentiment level*, they really do not need to be *domain specific*, but only *genre specific.* This reduces the number of models needed, as a syntactic tagging model for reviews might cover several domains, and thus the *domain independence* of the presented method is intact.

Especially the property of being domain independent is considered to be of significant importance of the presented method. As harvested data grows, and new domains surfaces (e.g. *Internet of Things*) any method requiring labeled training data will be slower and more costly to deploy. Besides the savings of avoiding expensive computational training of domain specific models, the method also allows the ability to *reuse* models on new and unseen domains, as long as some *domain expert* provides the seed concepts (S_{pos} and S_{neg}).

An interesting experiment would have been to see how the presented method performed on such genre specific syntactic models. [16] presents methods for *cross-domain semi-supervised learning*, i.e. the combination of labeled (e.g. *CCG-Bank*) and unlabeled (e.g. review texts) data from different domains (e.g. syntactic genre). This allows the construction of models that utilizes the knowledge present in the labeled data, but also biases it toward the distribution of the unlabeled data. The learning accuracy is of cause not as significant as compared to learning with large amounts of labeled target data, but it can improve genres where no labeled data are available.

7 Conclusions

This paper has presented a *formal logical method* for *entity level* sentiment analysis, which utilizes *machine learning techniques* for efficient syntactic tagging. The method should be seen as an alternative to pure machine learning methods, which have been argued inadequate for capturing long distance dependencies between an entity and opinions, and of being highly dependent on the domain of the sentiment analysis.

Empirical results showed that while the correctness of the presented method seems acceptably high, its robustness is currently inadequate for most real-world applications. However, it is argued that it indeed is possible to improve the robustness significantly with further refinements of the presented method.

Besides resolving the issue of the low robustness, the presented method also leaves plenty of opportunities for expansion. This could include a more sophisticated pronoun resolution, and even more advanced extraction strategies could also include relating entities by the use of some of the abstract topological relations available in semantic networks. E.g. *hyponym/hypernym* and *holonym/meronym*. With such relations, a strong sentiment of the entity *room* might inflict the sentiment value of *hotel*, since *room* is a meronym of *building*, and *hotel* is a hyponym of *building*.

In the future we aim to use advanced mathematical proof assistants like *Coq* (https://coq.inria.fr/) and *Isabelle* (https://isabelle.in.tum.de/) for the formalization of the presented theory. The proof assistants have support for the necessary data structures and algorithms. The use of proof assistants would allow for formal proofs of key properties and also for easier experiments with the presented method.

Appendix: Labeled Test Data

The following table consists of a random sample chosen from the "Swissotel Hotel" topic of the *Opinosis data set* [8] which contain any morphological form of the *subject of interest: hotel rooms*. Each sentence in the data set (which may not constitute a complete review) has been labeled independently by two human individuals *with respect to the subject of interest: hotel rooms*. Furthermore the table contains results for the presented method (entity level polarity value of subject of interest).

#	Review text	Humans	Method
1	The rooms are in pretty shabby condition, but they are clean	Negative	Unknown
2	The rooms are spacious and have nice views, I was NOT impressed with the mattress and every, little, tiny thing costs money	Unknown	N/A
3	The rooms look like they were just remodled and upgraded, there was an HD TV and a nice iHome docking station to put my iPod so I could set the alarm to wake up with my music instead of the radio	Positive	Unknown
4	The rooms were cleaned spic and span every day	Positive	Unknown
5	When I got to the room, I thought the new rooms would have a plasma since the website implies the new rooms would have them, but I guess those come later	Negative	Unknown
6	Very impressed with rooms and view!	Positive	Unknown
7	The rooms are not all that big	Negative	Unknown
8	Expensive Parking but great rooms	Positive	30.0
9	Rooms were nicely furnished	Positive	Unknown
10	The rooms are very clean, comfortable and spacious and up-to-date	Positive	52.0
11	I've only ever stayed in the "standard" rooms in this property, all of which are spacious and airy, and function well for both business or leisure travellers	Positive	Unknown
12	It does suffer, however, from a trend that I have been noticing that as rooms at business class hotels are upgraded, particularly with a patch panel for the big LCD, TV, drawer space becomes less and less	Negative	Unknown
13	We even got upgraded to one of the corner rooms which also looked west toward Michigan Ave and the Wrigley building	Positive	Unknown
14	The rooms were very clean, the service was polite and helpful, and it's near the heart of Chicago!	Positive	52.0
15	You can see downtown and or the Navy Pier from most of the rooms	Positive	Unknown
16	no more bathrobes in corner rooms suites, coffee service in room is parred way down, the buffet offered in the cafe is not as bountiful, although the cafe staff is impeccable and extremely gracious and will bring you what you wish, check in staff not at all eager to upgrade you, even though you may be a frequent visitor	Negative	Unknown
17	Our rooms were nice and didn't look worn or old	Positive	Unknown
18	Rooms at the hotel are getting somewhat tired	Negative	0.8

(Continued)

(Continued)

#	Review text	Humans	Method
19	Great Location great rooms and bed but no help from desk personnel	Positive	38.0
20	While the rooms are quite nice, I was dismayed by the snotty service I received at the Swissotel in Chicago	Positive	72.0
21	Rooms are dated, our corner room's bathroom was shabby	Negative	Unknown
22	The hotel was very nice, rooms were big, the pool hot tub area was very nice, and the location was great and easy to get to	Positive	10.0
23	Rooms are good quality and clean, what you would expect from a four star business hotel	Positive	46.0
24	The view from the rooms was fantastic, My daughters are allergic to feathers and all trace of them were removed from the room as soon as we advised housekeeping	Positive	Unknown
25	The Swissotel is one of our favorite hotels in Chicago and the corner rooms have the most fantastic views in the city	Positive	Unknown
26	Then again, the rooms are much larger and the view more than makes up for it	Positive	26.0
27	Rooms in similar hotels would usually be about $250, 300	Positive	Unknown
28	The actual hotel and rooms were very nice with amazing views, the staff was extremely rude	Positive	8.0
29	The rooms were clean, and upscale for the low price we paid	Positive	Unknown
30	Thanks to TravelZoo I was able to find an amazing deal, lakeside rooms for $129 night as part of a spring promotion	Positive	Unknown
31	I received a great deal on the rooms here and it was wonderful	Positive	8.0
32	The room was huge as hotel rooms go	Positive	26.0
33	Hotel was very clean and the rooms were comfy	Positive	Unknown
34	word to the wise, avoid the rooms ending with 11	Negative	Unknown
35	The rooms are large and well, appointed, the staff was very professional and friendly, and the view was striking!	Positive	34.0

References

1. Baldridge, J., Kruijff, G.J.M.: Multi-modal combinatory categorial grammar. In: Proceedings of the Tenth Conference on European Chapter of the Association for Computational Linguistics, EACL 2003, vol. 1, pp. 211–218 (2003)
2. Barendregt, H., Dekkers, W., Statman, R.: Lambda Calculus with Types. Cambridge University Press, Cambridge (2013)
3. Cambria, E., Schuller, B., Liu, B., Wang, H., Havasi, C.: Statistical approaches to concept-level sentiment analysis. IEEE Intell. Syst. **28**(3), 6–9 (2013)
4. Clark, S.: A supertagger for combinatory categorial grammar. In: International Workshop on Tree Adjoining Grammars and Related Frameworks, pp. 19–24, Venice, Italy (2002)
5. Clark, S., Curran, J.R.: Wide-coverage efficient statistical parsing with CCG and log-linear models. Comput. Linguist. **33**(4), 493–552 (2007)
6. Feldman, R.: Techniques and applications for sentiment analysis. Commun. ACM **56**(4), 82–89 (2013)
7. Fellbaum, C. (ed.): WordNet: An Electronic Lexical Database (Language, Speech, and Communication). The MIT Press, Cambridge (1998)
8. Ganesan, K., Zhai, C.X., Han, J.: Opinosis: a graph based approach to abstractive summarization of highly redundant opinions. In: International Conference on Computational Linguistics, pp. 340–348 (2010)
9. Hockenmaier, J.: Data and models for statistical parsing with combinatory categorial grammar. Ph.D. thesis, University of Edinburgh (2003)
10. Hockenmaier, J., Steedman, M.: CCGbank: a corpus of CCG derivations and dependency structures extracted from the Penn treebank. Comput. Linguist. **33**(3), 355–396 (2007)
11. Liu, B.: Web Data Mining: Exploring Hyperlinks, Contents, and Usage Data. Springer, Heidelberg (2007). https://doi.org/10.1007/978-3-642-19460-3
12. Marcus, M.P., Marcinkiewicz, M.A., Santorini, B.: Building a large annotated corpus of English: the Penn treebank. Comput. Linguist. **19**(2), 313–330 (1993)
13. Miller, G.A.: WordNet: a lexical database for English. Commun. ACM **38**(11), 39–41 (1995)
14. Pang, B., Lee, L.: Opinion mining and sentiment analysis. Found. Trends Inf. Retrieval **2**(1–2), 1–135 (2008)
15. Petersen, N.C., Villadsen, J.: Combining formal logic and machine learning for sentiment analysis. In: Andreasen, T., Christiansen, H., Cubero, J.-C., Raś, Z.W. (eds.) ISMIS 2014. LNCS (LNAI), vol. 8502, pp. 375–384. Springer, Cham (2014). https://doi.org/10.1007/978-3-319-08326-1_38
16. Søgaard, A.: Semi-supervised learning. In: ESSLLI-2012 Lecture (2012)
17. Simančík, F., Lee, M.: A CCG-based system for valence shifting for sentiment analysis. Res. Comput. Sci. **41**, 99–108 (2009)
18. Steedman, M.: The Syntactic Process. The MIT Press, Cambridge (2000)
19. Steedman, M.: Taking Scope: The Natural Semantics of Quantifiers. The MIT Press, Cambridge (2011)
20. Tan, L.K.-W., Na, J.-C., Theng, Y.-L., Chang, K.: Sentence-level sentiment polarity classification using a linguistic approach. In: Xing, C., Crestani, F., Rauber, A. (eds.) ICADL 2011. LNCS, vol. 7008, pp. 77–87. Springer, Heidelberg (2011). https://doi.org/10.1007/978-3-642-24826-9_13

Reforming AMR

Edward Stabler[1,2](✉)

[1] University of California, Los Angeles, CA, USA
[2] Nuance Communications, Sunnyvale, CA, USA
edward.stabler@nuance.com

Abstract. Many recent proposals aim to simplify semantic representations, and Abstract Meaning Representation (AMR) comes from this tradition, but it is nevertheless quite expressive. Bos 2016 proposes a slightly reformed AMR for translation to first order logic. This paper proposes a different augmentation of AMR that is more easily provided, and a slightly different mapping to higher order and dynamic logic. The proposed augmentation can be, at least in most cases, easily computed from standard 'unreformed' AMR corpora. The mapping from this augmented AMR to logical representation is a finite state multi bottom up tree transduction.

With a variety of scientific and engineering motivations, a number of recent studies have aimed to simplify semantic representations in various ways that include reducing recursion, reducing the number of modes of composition, and leaving some ambiguities unresolved [2,14,36,40]. Abstract Meaning Representation (AMR) [4,5,27] falls into this tradition; it was initially designed to be easily learnable by automated translation systems [31]. An ongoing effort uses AMR to annotate large corpora [27] for engineering applications and other quantitative studies of the important question: which constructions are used to mean which things in which contexts?

In [4], the AMR on the left below is proposed to represent the meaning of *The boy wants to go.* The AMR on the right is slightly augmented with syntactic features in the leaves and a :quant arc for the determiner *the*, as discussed below.

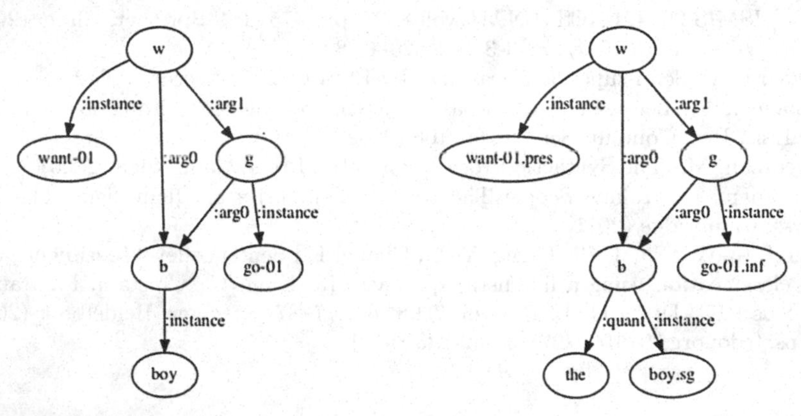

A. Foret et al. (Eds.): FG 2017, LNCS 10686, pp. 72–87, 2018.
https://doi.org/10.1007/978-3-662-56343-4_5

In the graph on the left, above, the 3 leaves (i.e., the nodes with no outgoing arcs) are disambiguated lexical concepts, while the 3 internal nodes are variables, written with a single letter optionally followed by a number. The 6 arcs are labeled with roles written with an initial colon. Intuitively, the graph on the left could be read: "there is an :instance of wanting, w, with an :arg0 (subject) role filled by an instance b of a boy, and with an :arg1 (theme) role filled by an :instance g of going, which in turn has an :arg0 role again filled by b." In this version of AMR, the arc labels and the relevant senses of the verbs *want* and *go* are taken from OntoNotes [41] and PropBank [8].

So AMR indicates verb senses, coreference, and some aspects of argument and modifier structure. In some discourse contexts, the prepositional phrase modifier in *The money was stolen by the bank* is an agent and not a locative. And in some contexts, a spoken or casually written *let's meet at last call* could intend the prepositional phrase to specify not a time but a location, one of the many bars named 'Last Call'. In a large corpus, some of the relevant cues for such distinctions can be studied, cues to which normal human speakers are obviously exquisitely attuned. But standard AMR does not indicate tense, plurality, quantification, or scope. For some engineering applications, tense, plurality and quantification may not matter, but for other applications it is obviously important to get the whole message of an utterance approximately right. Following preliminary work by Bos and others [3,6,9], this paper argues that with a very minor reform that does not significantly affect the syntactic complexity of AMR notation, it can become considerably more sophisticated. The reform is indicated in the augmented AMR (AAMR) shown above on the right: we add tense to leaves corresponding to verbs, grammatical number to leaves corresponding to nouns, and for each noun concept, if that noun has an associated article (*the, a*) or quantifier (*every, some, most, exactly 5, between 10 and 20, ...*), those go into the :quant role.

1 AMR Triples and Trees

Each AMR is a connected, directed, arc-labeled graph with a designated 'root' node, where (a) at most one node has no incoming arcs, and when there is such a node, it is the designated 'root', (b) any node with an outgoing arc is a variable, (c) the arc-labels are roles (chosen from a small, finite set), (d) every internal node has a unique outgoing arc with the :instance role, and (e) every leaf is a lexical concept, where a lexical concept is a sense-disambiguated stem possibly with tense, number, gender information. We can represent any AMR by a pair (rootNode, arcs) where arcs is a set of (sourceNode,role,targetNode) triples. So for example the AAMR displayed on the right above is:

(w, { (w,:instance,want-01.pres), (w,:arg0,b), (w,:arg1,g),
(b,:instance,boy.sg), (b,:quant,the),
(g,:instance,go-01.sg), (g,:arg0,b) })

Note that in this AAMR, one and the same :instance of a boy b plays a role in two predications, because *want* is a 'control verb' [29]. English obviously has many ways to indicate intended coreference. When we construct AAMRs for conjunctions of sentences and discourses, it is common to have many long chains of coreferring argument expressions. The following is a very simple case, showing again a notation on the right that is discussed below.

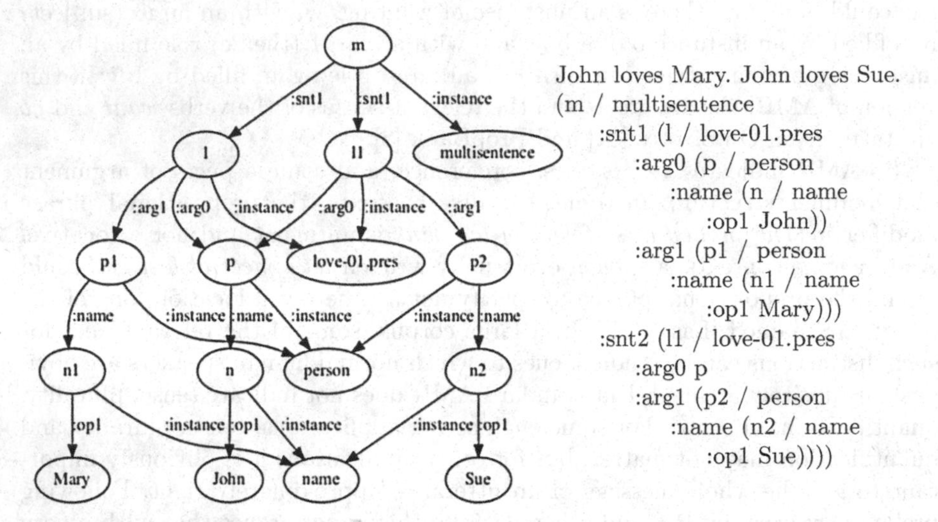

In this structure we see that the named entities are classified by lexical concepts (like *person*) that do not correspond to explicit items in the sentence, as explained in [5]. This classification of named entities will not be a focus in this paper.

[4, Sect. 2] says AMRs are rooted and acyclic, but [5] provides a non-rooted, cyclic structure for *a procedure to ensure quality*, which we also allow, augmented:

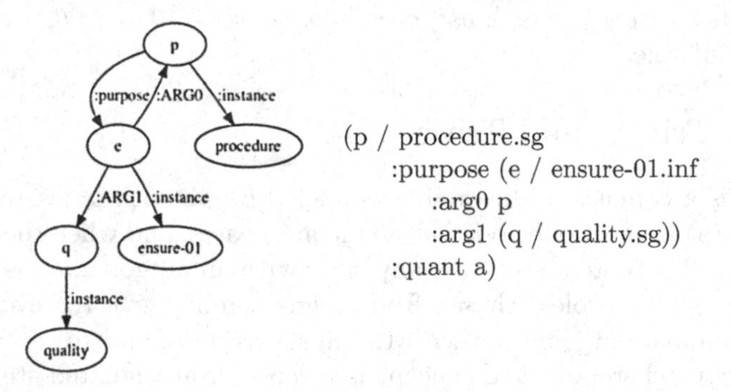

[4, Sect. 2] uses a node labeling function, which we do not need for AMRs, but only for the AMR tree representations introduced in this section, as discussed above. And in addition to lexical concepts, [5] uses some special constants, like '-' on a :polarity arc to indicate negation. But '-' can at least sometimes be regarded as the representation the lexical 'not'; see Sects. 2 and 3 below.

Letting the K-width of a directed graph be the maximum number of distinct simple paths (possibly overlapping) between any two nodes [20], it is easy to see that there is no finite bound on the K-width of AMR structures of English.

For example, even just with arbitrarily long conjunctions of the form *John loves Mary and John loves Sue and...*, in which the love relation can be asserted between any pair of named individuals in any finite set, we can make K-width arbitrarily high. Notice that even as K-width grows quite high, discourses with this kind of structure (perhaps adding some modifiers) can be not only intelligible but exciting in some contexts. Ignoring the direction of the arcs, AMR structures for English sentences also have unbounded treewidth [1,15]; any number of nodes in these love relations can be connected to each other and involved in any number of cycles.[1] AMRs are not simple graphs. But for some calculations, we can split the AMR nodes with more than one predecessor to make a tree, putting all descendants of the split node under only one of its copies, and calculate on the tree without ever needing to check the identities among those split nodes. All the calculations described in this paper have that property.

Turning to the tree representations for AMRs already shown above, these can be derived from AMRs by the introduction of a node labeling function which is the identity function for every node with $p \leq 1$ predecessors, but for any node n with $p > 1$ predecessors in the graph, the tree has p different nodes $\mathrm{n}_0, \ldots, \mathrm{n}_{p-1}$ all labeled n, with all descendants of the original node attached under only one of the copies n_i in the tree. As we saw above, because AMRs can be cyclic, it can happen that no node lacks a predecessor, but when that happens the 'designated root' is chosen as the root of the corresponding tree. Applying these ideas to the AAMR on the first page of this paper, and abbreviating :instance with / (as is standard in the literature), we obtain a tree that can be drawn or pretty-printed:

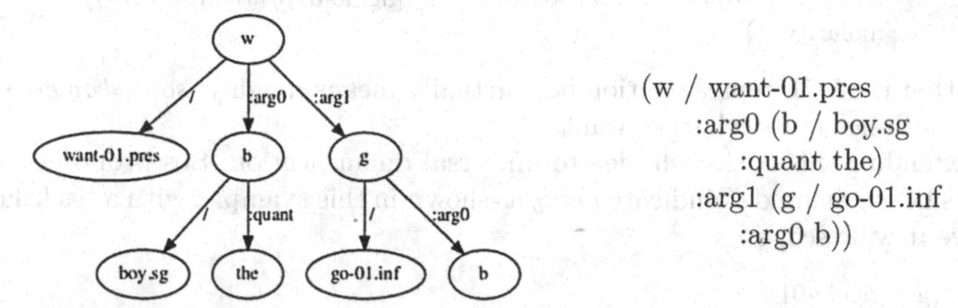

Additional examples of the pretty-printed tree format have already been provided for the love and purpose-clause examples above.

In order to have a unique tree representation for each AMR, let's initially adopt the convention that (i) subtrees are ordered left-to-right with the :instance subtree (often labeled /) first and otherwise in standard alphanumeric order, and

[1] A tree decomposition of undirected graph $\mathrm{g} = (\mathrm{V},\mathrm{R})$ is a tree $\mathrm{t} = (\mathrm{U},\mathrm{S})$ where (i) $\bigcup_{\mathrm{u} \in \mathrm{U}} \mathrm{u} = \mathrm{V}$, (ii) if an arc in g connects v_i and v_j, then some $\mathrm{u} \in \mathrm{U}$ contains both v_i and v_j, and (iii) if some node v of g is in two nodes $\mathrm{u}_i, \mathrm{u}_j$ of U, then v is in every node on the path between u_i and u_j. The treewidth of a decomposition is $\max_{\mathrm{u} \in \mathrm{U}} |\mathrm{u}| - 1$. The treewidth of g is the minimum treewidth over all decompositions of g. Many problems have complexities that increase with treewidth [7,20], and Courcelle's theorem relates treewidth to MSO definability [15]. Computing treewidth is NP-complete, but code for computing treewidth of small graphs is available at [1].

(ii) for any set of nodes with the same label, all must be leaves except possibly the one that occurs first in the preorder traversal. (i) is revised in Sect. 3 below.

2 Raising Negation and First Order Quantifiers

The intuitive reading of AMRs suggested in the first paragraph of this paper binds the variables of an AMR with existential operators. But that will not work when negation is attached under the verb. As in some dependency grammars. [9] proposes a translation from AMRs with negation to first order logic (FOL), raising the negative polarity to take wide scope. Assuming *The boy didn't giggle* gets the AMR on the left, it is mapped to the FOL translation on the right:

```
(g / giggle-01
    :arg0 (b / boy)
    :polarity - )
```

$$\neg\exists g(\text{giggle-01}(g) \wedge \exists b(\text{boy.sg}(b) \wedge \text{:arg0}(g,b)))$$

But that FOL means something like *No boy giggled*, while the sentence *The boy didn't giggle* is clearly talking about some particular boy. For this, Bos proposes a reform of AMR which marks constituents with a backwards :instance slash whenever they should be raised to take scope over negation. In the translation to FOL, a concept :instance introduced with the backward slash takes scope over the negation:

```
(g / giggle-01
    :arg0 (b \ boy)
    :polarity - )
```

$$\exists b(\text{boy}(b) \wedge \neg\exists g(\text{giggle-01}(g) \wedge \text{:arg0}(g,b)))$$

Note that the FOL representation here actually means roughly: *some boy doesn't giggle*, still not quite what we want.

Extending this backslash idea to universal quantification, Bos proposes that AMRs be augmented to indicate *every* as shown in this example, with a backslash to give it wide scope:

```
(g / giggle-01
    :arg0 (b \ boy
              :quant A ))
```

$$\forall b(\text{boy}(b) \rightarrow \exists g(\text{giggle-01}(g) \wedge \text{:arg0}(g,b)))$$

In summary, Bos's proposal is roughly this: (1) The event structure associated with each verb is existentially closed; (2) negation scopes over the existential closure of the event that the :polarity role is attached to; (3) arguments marked by reversing their :instance slash are raised to take scope over the whole structure; and (4) alternative permutations of the quantifiers with reversed slashes are generated nondeterministically.

These steps head in the right direction, but observe these four points:

- If articles, grammatical number and tense are not indicated, then the AMR assigned to Bos's example *The boy didn't giggle* is also assigned to *boys don't giggle* and *the boys will not giggle* and *a boy won't giggle*. These sentences mean different things in respects that could be important for some applications of AMR annotation, and it is easy to add a little more to the annotation to distinguish them, as proposed in the introduction to this paper.

- Bos's backslash proposal (3,4) provides a way to allow arguments to scope over the existential event quantifier, but, as noted for example by Champollion [12] and Landman [30], it is not the exception but the rule that this should happen. For various theoretical and empirical reasons, it is more natural to take a sentence like *John kissed every girl* to assert possibly many different kissing events, one for each girl, rather than a single event that includes possibly many different girls at many different times and places. And Bos points out that among the phrases that regularly outscope the event quantifier and negation are not only quantified phrases like *every boy* but also "proper names, appositive expressions, definite descriptions, and possessive constructions." This could be built into the translation, rather than requiring every instance to be marked. This idea is developed in Sect. 3.
- When AMRs are used to represent conjunctions of sentences and discourses, proposals (3,4) will raise quantifiers over the whole conjunction or discourse, and permuting them will generate spurious scope interactions. Scope inversions are usually clause- and sentence-bound, and even with those restrictions, the number of alternative scopings can grow quickly. Scope locality restrictions are imposed in Sect. 3, and we don't generate alternatives.
- As noted above, since the definite article is ignored in current AMR, the FOL translation for *the boy didn't giggle* actually means something like *some boy didn't giggle*. Articles have not only quantificational force but also a discourse role that affects utterance meaning. AAMRs, augmented with quantifiers and articles as suggested in the introduction, allow a translation to logic that can respect the force of these elements. This idea is briefly developed in Sect. 5.

3 Towards a Proper Treatment of Quantifiers

The survey in [38, Sects. 13, 14, 15 and 16] shows that among the many things not expressible by any iterations of first order quantifiers are the following, using As, Bs for determiner phrase denotations, R for properties/relations, and R-er for comparatives: *most As R, more than one third of As R, 80% of As R, more As than Bs R, an even number of As R, different As R different Bs, As are usually R-er than Bs, an infinite number of As R*. A longer, wider-ranging list is provided in [26]. And the list gets much longer if we include quantifiers whose representation in FOL with identity is possible but not feasible: it is an old point that FOL with identity is not a good language to express propositions like *More than seven billion people need clean water*. FOL is not appropriate for linguistic semantics. And even for a wide range of common engineering applications, operators for cardinalities and cardinal comparisons are common enough that logics like OWL, carefully designed for engineering goals, provide them [7,24].

So we begin by proposing a translation from AAMR to higher order logic (HOL) in which quantifiers like *every* and *some* are treated like *most, seven billion, more than 1/3 of,...*. Let's assume that these quantifiers apply to pairs of properties to yield truth values. (We revise this assumption about the higher order type of binary quantifiers in section Sect. 5.) For example, taking a simple variant of Bos's example, we will map the AAMR for *Most boys giggled* on the left to this HOL on the right:[2]

```
(g / giggle-01.pres
   :arg0 (b / boy.pl                most(boy.pl,λb∃g(giggle-01.pres(g) ∧ :arg0(g,b)))
            :quant: most ))
```

That is, roughly, most boys have the property that there is an event in which they giggle. Note that we scope *most* over the event existential by default, rather than needing a backslash to trigger that effect.

Adding negation to get *Most boys do not giggle*, there are actually two readings: either the 'surface order' (i) most boys are such that it's not the case that they giggle, or the 'inverted order' (ii) it's not the case that most boys are such that they giggle. This is a case where the two readings are not so easy to disassociate. But if we assume that *most* means more that half, precisely, then in a situation where exactly half the boys giggle, then (i) is false, but (ii) is true. As is often the case in structures with interacting scopes, I think the surface order reading is the most natural here:

```
(g / giggle-01.pres
  :arg0 (b / boy.pl
           :quant most )          most(boy.pl,λb¬∃g(giggle-01.pres(g) ∧ :arg0(g,b)))
  :polarity - )
```

Rather than generating the non-surface order scopal orders for AAMRs, we will generate just the basic surface order reading. If needed, alternative orders can be computed from the basic order either with an algorithm that raises the quantifiers to get all possible alternative orders, or by using some kind of heuristic method to generate only scopes that are most likely in some sense, as discussed in Sect. 4.

In sum, while Bos computes all the raised quantifiers for the whole structure together, and then applies them to the representation of the unraised structure, we will only raise quantifier arguments to scope over the AAMR that they are arguments of, to generate surface-order scope only. We can see already that this calculation of HOL from the tree representation of AAMR is quite simple; it can be done by a simple finite state mechanism, a deterministic multi bottom up tree transducer (mbutt) [16,19]. To set the stage for that, first we revise our earlier convention about the linear order in tree representations of AAMRs: instead of putting subtrees into alphanumeric order, we will order them according to the

[2] HOL with generalized quantifiers is introduced, for example, in Carpenter's [11, Sect. 3]. Like Carpenter, we write $\forall x \phi$ for every$(\lambda x.\phi)$, and $\exists x \phi$ similarly.

linear order of their corresponding elements in the input string.[3] Second, we will modify the tree representation so that the roles are not treated as arc labels but as node labels. This allows AMRs to have a standard term representation. So our previous example actually is given to the transducer as the tree on the left:

```
g(:instance(giggle-01.pres),
  :arg0(b(:instance(boy.pl),
          :quant(most))),          most(boy.pl,λb¬∃g(giggle-01.pres(g) ∧ :arg0(g,b)))
  :polarity(-))
```

Intuitively, a bottom up tree transducer differs from a bottom up tree acceptor in that the states may have a subtree, their output. An mbutt differs from a simple tree transducer in allowing states to have more than one subtree, so that up to k subtrees can be carried up the tree. What needs to happen in our transformation is that the subtrees corresponding to negation and to each argument of each lexical concept need to be lifted to scope over each verbal concept. We will define the transducer so that the order of the quantifier subtrees is preserved as they raise, so that the transducer can place those elements in the order that corresponds to the string surface order.

Slightly more formally, a bottom up tree acceptor A $= (Q, F, \Sigma, \delta)$ where Q is a finite set of states, $F \subseteq Q$ is the set of final states, Σ is a ranked alphabet, and δ is a finite set of transition rules of the form $f(q_1, \ldots, q_n) \to q$, mapping a tuple of states to a state. Leaves are 0-ary, and hence treated by rules of the form $f \to q$, mapping a leaf to a state. The acceptor is deterministic iff no two rules have the same left side. To get a transducer we let the states be 1-ary, with rules of the form $f(q_1(t_1), \ldots, q_n(t_n)) \to q(t)$, where t may contain $t_1, \ldots, t_n$ as subtrees, together with any fixed structure. If a rule has at most one occurrence of each t_i in t, it is linear, and the transducer is linear iff all its transitions are. And we extend this to mbutts by letting the states have arity up to k for some k. See [16,18,19].

Formalizing and implementing the transduction we need for any corpus is straightforward, but slightly tedious since many elements are rearranged in the HOL. Stepping through the simple example above, *most boys do not giggle*, will make clear what the task requires. Since HOL allows higher order functions, the expressions denoting functions can be complex, and so expressions like (f(g))(h) are well formed – with some operator precedence conventions this is written more simply as fgh. To represent such expressions as trees, we will make the applications explicit, using a period as the application operator, representing (f(g))(h) with the tree .(.(f,g),h):

[3] The correspondence between AAMR subgraphs and elements of the input string is sometimes given by hand-specified alignments, and there are a number of proposals about how to compute them when hand-specifications are not available [13,17]. Note that ":polarity -" will be aligned with the negation in the input string. In the example above with 'most' and 'not', the surface order and the alphanumeric order coincide.

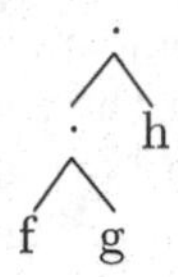

Since the translation from (f(g))(h) to .(.(f,g),h) is easy and the former is easier to read, in the following discussion we use the former notation, sometimes omitting parentheses when no confusion results. And sometimes, $\lambda x.\mathrm{p}(x)$ is reduced to p.

We assume that the set of basic vocabulary of concepts w is fixed and given. For every concept w, we have a rule mapping that concept to a 1-ary state q_c that has the concept as an argument:

$$\mathrm{w} \rightarrow q_c(\mathrm{w})$$

That is, for the leaves of the AAMR just above for *most boys do not giggle*, we have the rules:

$$\begin{array}{l} \text{giggle-01.pres} \rightarrow q_c(\text{giggle-01.pres}) \\ \text{boy.pl} \rightarrow q_c(\text{boy.pl}) \\ \text{most} \rightarrow q_c(\text{most}). \\ \text{-} \rightarrow q_c(\text{-}) \end{array}$$

Now moving up the tree, we have the :instance role dominating the state q_c(giggle-01.pres). For these :instance roles, we have the rule

$$\text{:instance}(q_c(t)) \rightarrow q_i(t).$$

So at that node, we now have the state q_i(giggle-01.pres). Climbing up from -, we have the :polarity role, a special case handled by the rule

$$\text{:polarity}(q_c(\text{-})) \rightarrow q_{-}.$$

Climbing up from boy.pl use the instance rule just above to get q_i(boy.pl). Climbing from most, we have the :quant role and use the rule:

$$\text{:quant}(q_c(t)) \rightarrow q_q(t).$$

Now we get to the more interesting steps. The AAMR dominated by b is processed with this rule:

$$\mathrm{b}(q_i(t_0), q_q(t_1)) \rightarrow q_{qa}(t_1, \mathrm{b}, t_0).$$

So this last step yields q_{qa}(most,b,boy.pl). Climbing from that last step, the :arg0 role is assigned:

$$\text{:arg0}(q_{qa}(t_0, t_1, t_2)) \rightarrow q_{rcq}(\text{:arg0}, t_0, t_1, t_2).$$

At this point we have q_{rcq}(:arg0,most,b,boy.pl). Now all the needed parts are available for the last step:

$$\mathrm{g}(q_i(t_0), q_{-}, q_{rcq}(t_1, t_2, t_3, t_4)) \rightarrow q_a(t_2(t_4, \lambda t_3 \neg \exists \mathrm{g}(t_0(\mathrm{g}) \wedge t_1(\mathrm{g}, t_3)))).$$

This rule yields the desired q_a(most(boy.pl, λb$\neg\exists$g(giggle-01.pres(g) $\wedge$:arg0 (g, b)))).

Clearly, this kind of deterministic bottom up assembly of HOL is going to be possible as long as the number of subtrees needed at each step is finite and the number of different things we need to do with those subtrees – controlled by the states – is finite. AMR does not impose a fixed bound on the number of argument roles a predicate can have, but for any fixed AMR corpus, there will be a maximum. And we carry quantifiers only up to the verbal sub-AMR that introduces them – a kind of locality restriction on scope interaction – so we only ever have a finite number. So clearly, within the finite bounds needed for any corpus, an mbutt mapping AAMR to the surface-order scope HOL can be defined. This finite state transduction is ϵ-free and deterministic, but not linear because of variable copying.

4 Bounded Quantifier Raising

The calculation described in the previous section has nice properties, but it leaves quantifiers in their low 'surface order' positions. If we provide a fixed finite bound on the number of quantifiers requiring non-surface scope, a nondeterministic mbutt can be defined that permutes the quantifiers to get the alternative scopes. We sketch an extension to the transducer of the previous section with this effect, but a similar alternative approach could define the quantifier-raising transducer independently, and compose it with the previous transducer when alternative scopes are desired.

Hobbs and Shieber [23] observe that the following sentence with 3 quantifiers has not 3 != 6 but only 5 alternative scopes:

Every representative of some company saw most samples.

The missing reading is the one where *every representative* takes widest scope, and *most samples* scopes over *some company.* That scope order is excluded because *representative* is relational and the *of*-phrase names one of its arguments.[4] Consider first the HOL surface scope of these simpler sentences:

Every representative saw most samples
surface scope: every(representative, λy.most(sample, λx.saw(x, y)))

Every representative of some company laughed
surface scope: every(λx.some(company, λy.representative(x, y)), laughed)

[4] The discussion in Hobbs and Shieber has an error that does not affect their main point. Their example sentence is not talking about things that are both representatives and also of-some-company – that doesn't quite make sense intuitively, and in fact gets the wrong entailments; see e.g. [37]. Rather, *representative* is relational and *of some company* specifies one of its arguments. We rephrase the Hobbs and Shieber argument here without that mistake. In the LDC AMR corpus [27], *representative* is treated relationally as it should be, as denoting an :arg0-of the predicate represent-01, where :arg0 is the representer and :arg1 is the thing represented.

Note that in the former example, *most* is in the second argument (sometimes called the 'scope') of *every*, while in the latter example, *some* is in the first argument (the 'restriction') of *every*. Returning now to the sentence considered by Hobbs and Shieber, it is clear that we cannot put *most* between *every* and *some*, since *some* must be in the first argument of *every*, and *most* needs to be in the second argument of *every*. Hobbs and Shieber mention a number of quantifier-scoping proposals that derived ill-formed structures for the missing reading, and they also note that while reducing 5 to 4 may seem unimportant, the same consideration reduces the possibilities for the following sentence from 5 != 120 to 42:

> Some representative of every department in most companies saw a few samples of each product.

The ill-formed structures are avoided by any approach that respects the binary type of these quantifiers and only raises them, never lowering a quantifier into the argument of another quantifier.

It is now easy to see that a nondeterministic mbutt can do this quantifier raising if there is a fixed, finite bound on the number of quantifiers to raise. For the previous 2 simple examples, consider what is required to derive these non-surface scopes:

Every representative saw most samples
inverse scope: most(sample, λy.every(representative, λx.saw(x, y)))

Every representative of some company laughed
inverse scope: some(company, λy.every(λx.representative(x, y), laughed))

The former inverse LF means something like: *most samples are in the set of things that every representative saw*, and the latter inverse LF means roughly: *some company is such that every one of its representatives laughed.* To derive the former LF, instead of using the rule placing *most* and its restriction into surface scope position, these are lifted up to take wide scope over the whole formula. In the latter case, instead of using the rule placing *some* and its restriction into surface position, we lift them to wide scope. Comparing the surface and inverse representations for each of these examples, notice that just finitely many subtrees are lifted in each case – and they are, in fact, subtrees already computed by the transduction from AMR to surface order HOL. Clearly, with a finite bound on the number of moving quantifiers, we can define rules to place the moving quantifiers in all possible orders.

The quantifier raising steps are nondeterministic but it is ϵ-free and linear; as we see in the examples above, the quantifier raising step rearranges subtrees from the surface order, and never needs to delete or produce multiple copies of those subtrees. The Hobbs and Shieber algorithm achieves the same effect, but one advantage of a linear mbutt representation is that linear mbutts are closed with respect to composition [16,19]. For example, we can compose an mbutt that accepts any particular input HOL with the quantifier raising transduction to get a representation of the whole set of alternative scopes for the input.

Hobbs and Shieber point out that many preferences can be incorporated relatively easily by controlling the nondeterministic choices, and at least some of these appear to be within the range of weighted mbutts [33] – a topic for future work.

There is an enormous and still growing literature on quantifier scoping patterns and preferences. In a recent summary review, Szabolcsi [39] discusses some preferences that were not mentioned in the paper by Hobbs and Shieber more than 20 years earlier, but points out that the main advance is the recognition that many different things are interacting to produce effects that are sometimes misleadingly lumped together as scope interactions. We split out one of the important factors in Sect. 5.

5 Dynamics and Accommodation

We conclude with a brief sketch of a possible approach to the definite article and some of the many things related to that.[5] As noted at the beginning of this paper, when an entity is the argument of more than one predicate, roles from the respective predicates point to a single representation of that entity. This can happen in control constructions, with pronouns, with proper names, and with descriptions. Computing the coreference relations as a typical speaker would is clearly a hard problem [35], so as in the previous section, we consider how to factor this hard problem away from other aspects of AAMR calculation.

In a sense, the coreference in the first example of the paper is the easiest to recognize: it is unambiguously determined by the grammar of English control constructions. Repetition of names in a discourse usually, but not always, signals coreference, as in our second example. Recognizing intended coreference in descriptions and pronouns is obviously harder:

John giggled. He's delighted.
John giggled. The boy is very happy.

Though the details remain controversial, there are clearly common elements in these coreference determinations, whether they involve control verbs, names, pronouns or descriptions. Just as *The present king of France is bald* presupposes that the subject describes something, so *He is bald* presupposes that there is something salient and male available to refer to. To handle this in discourse requires some notion of context that extends across sentence boundaries. This can seem impossible because context includes all kinds of things and grows without bound, but Montague famously pointed out that it is not necessary to consider contexts in their full complexity: we need only track what is essential for understanding

[5] Here we focus on the definite article, sketching briefly the fundamental change to a dynamic perspective. But indefinite articles are even trickier and complicate the picture of how scope works, impeding progress until it was recognized that they require special treatment. As discussed for example in Kratzer [28] and references cited there, they are unlike quantifiers like *every* or *three*, unlike referential expressions formed with *the*, and not adequately handled by the discourse closure proposed by Heim [22] and DRT [25]. See e.g. [10].

the particular discourse [34]. Furthermore, we have the fact that humans can not only do this with little or no conscious effort, but can predict fairly well how others will do it in a given context, as any careful writer knows.

There is a large literature about coreference resolution and tracking entities in discourse, but here I will sketch the bare outlines of a preliminary framework proposed by Lebedeva [32], based on de Groote [21]. Where the static theory has propositions of type o, this framework gives them the type $\gamma \to (\gamma \to o) \to o$, a map from contexts γ to continuations $\gamma \to o$ to propositions o. This 'dynamization of types' can be propagated throughout the type hierarchy. The logical constants are also dynamized so that, for example, conjunction applies the update of the first proposition and then the update of the second. And the terms are similarly dynamized, with a special treatment reserved for pronouns and the definite article. The selection of an antecedent for a pronoun can be modeled by assuming the existence of a dynamic selection function $\widetilde{\mathsf{sel}}$ which takes the context to return a dynamic entity. Let $\widetilde{\text{it}}$ represent the dynamized term that represents a function that selects a referent from context. Similarly let $\widetilde{\text{the}}$ represent a function that combines with a dynamic property to return an entity from the discourse context. When appropriate entities are not present, these functions can raise an exception that triggers an accommodation, e.g. the listener could just add a relevant entity. This is a big picture with many parts, but here I just want to make a small suggestion.

Consider again the simple example *The boy giggled.* In AAMR, the past tense and the definite article are indicated, and so now the natural proposal for its HOL representation is this:

```
(g / giggle-01.past
   :arg0 (b / boy.pl                 the(boy.pl,λb∃g(giggle-01.pres(g) ∧ :arg0(g,b)))
            :quant: the ))
```

Obviously, adding this treatment of the definite article has no effect on the complexity of the AAMR to HOL mapping. This approach simply marks this sentence as unlike the dynamized representation of *Some boy giggled* in a relevant respect. The operator $\widetilde{\text{the}}$ signals that an appropriate boy should be selected from context, or if that's not possible, some kind of accommodation should be triggered. Lebedeva proposes that not only definite descriptions, but names and pronouns should be treated in an analogous fashion.

Recognizing that the calculation of the selection function $\widetilde{\mathsf{sel}}$ is often challenging, notice that we could first calculate AAMR without the coreference links, without evaluating $\widetilde{\mathsf{sel}}$. In the example 2-sentence discourse given in Sect. 1, that would mean having two instances of individuals named *John.* Computing AMRs without selection is obviously going to be easier in many cases. Instead of an HOL translation of the AAMR given for the small dialog in Sect. 1, with the two occurrences linked, we could instead have only two dynamic terms. Elements that look for antecedents in the context can be marked in the translation but left unresolved. Coreference resolution, application of selection function, can then be done in a separate step and evaluated separately to see whether it assumes the same coreference links as a human speaker probably would in the same context.

6 Reforming the Corpus and Other Future Work

Why not translate directly from the strings in context to the higher order dynamic logic, dispensing with the AAMR? **First**, some practical points: AMR already annotates fairly large corpora, and it is being developed largely by volunteers and academics who disagree about what counts as an appropriate annotation for linguistic meaning. For this reason, a shared effort necessarily focuses on fundamental points of agreement. And for this same reason, it is relevant that the revisions proposed here are minor. AAMR adds some additional lexical information to AMR leaves, and attaches quantifiers to the nominal concepts they are associated with. In an aligned and parsed AMR corpus, the generation of the corresponding AAMR can be done largely mechanically. There is also a **second**, more elusive but possibly more important motivation for AMR, noted by many of the projects aiming to simplify semantic representations. For both practical and scientific reasons, it can be useful to have a semantic formalism that makes it easy to express those things that are most commonly expressed in natural language, without requiring a precision in interpretation that is completely unlike anything humans do. AMR is designed to fit the language, composing verb frames with classified argument and modifier roles, without settling the sometimes complex issues about quantifier scope, etc.

AAMR specifies a number of things that standard parse trees do not explicitly provide: verb senses, a semantic classification of arguments and modifiers, and intended coreference relations. This paper shows that fairly sophisticated logical representations with higher order argument structure can also be computed from these structures. If the computation of coreference is left to a later step, AAMR calculation should be quite feasible when context makes argument and modifier roles clear. And if determining the intended scope of quantifiers is left to a later step, the translation from AAMR into HOL is easy to define and compute. We have shown that when scope is local and scope determination is postponed, the translation to HOL is finite state and deterministic. Additional lexical information in the leaves sets the stage for treating other semantically important matters, some of which may similarly allow some of the relatively tractable parts of meaning representation to be factored away from the harder and less understood parts. In at least many cases, this approach will produce satisfactory results. In future work we hope to contribute to the production of an AAMR corpus with transducers that map it to reasonable HOL.

Acknowledgments. Many thanks to the anonymous reviewers for their valuable suggestions.

References

1. Adler, I., Dell, H., Husfeldt, T., Larisch, L., Salfelder, F.: The parameterized algorithms and computational experiments challenge - Track A: Treewidth (2016). https://pacechallenge.wordpress.com/pace-2016/track-a-treewidth/
2. Allen, J.F., Swift, M., de Beaumont, W.: Deep semantic analysis of text. In: ACL, SIGSEM Symposium on Semantics in Systems for Text Processing (STEP) (2008)
3. Artzi, Y., Lee, K., Zettlemoyer, L.: Broad-coverage CCG semantic parsing with AMR. In: 2015 Conference on Empirical Methods in Natural Language Processing, pp. 1699–1710 (2015)
4. Banarescu, L., Bonial, C., Cai, S., Georgescu, M., Griffitt, K., Hermjakob, U., Knight, K., Koehn, P., Palmer, M., Schneider, N.: Abstract meaning representation for sembanking. In: Proceedings of the 7th Linguistic Annotation Workshop and Interoperability with Discourse, pp. 178–186 (2013)
5. Banarescu, L., Bonial, C., Cai, S., Georgescu, M., Griffitt, K., Hermjakob, U., Knight, K., Koehn, P., Palmer, M., Schneider, N.: Abstract meaning representation 1.2.2 specification (2015). https://github.com/amrisi/amr-guidelines/blob/master/amr.md. Accessed 18 Sept 2015
6. Bender, E.M., Flickinger, D., Oepen, S., Packard, W., Copestake, A.: Layers of interpretation: on grammar and compositionality. In: 11th International Conference on Computational Semantics, pp. 239–249 (2015)
7. Bienvenu, M., Kikot, S., Podolskii, V.: Tree-like queries in OWL 2 QL: succinctness and complexity results (2015). https://arxiv.org/abs/1406.3047
8. Bonial, C., Babko-Malaya, O., Choi, J.D., Hwang, J., Palmer, M.: http://clear.colorado.edu/compsem/documents/propbank_guidelines.pdf (2010)
9. Bos, J.: Expressive power of abstract meaning representations. Comput. Linguit. **42**(3), 527–535 (2016)
10. Brasoveanu, A., Farkas, D.F.: How indefinites choose their scope. Linguist. Philos. **34**, 1–55 (2011)
11. Carpenter, B.: Type-Logical Semantics. MIT Press, Cambridge (1997)
12. Champollion, L.: The interaction of compositional semantics and event semantics. Linguist. Philos. **38**(1), 31–66 (2015)
13. Chen, W.T., Palmer, M.: Unsupervised AMR-dependency parse alignment. In: 15th Conference of the European Chapter of the Association for Computational Linguistics (2017)
14. Copestake, A., Flickinger, D., Pollard, C., Sag, I.: Minimal recursion semantics: an introduction. Res. Lang. Comput. **3**, 281–332 (2005)
15. Courcelle, B., Engelfriet, J.: Graph Structure and Monadic Second-Order Logic. Cambridge University Press, New York (2012)
16. Engelfriet, J., Lilin, E., Maletti, A.: Composition and decomposition of extended multi bottom-up tree transducers. Acta Informatica **46**(8), 561–590 (2009)
17. Flanigan, J., Thomson, S., Carbonell, J., Dyer, C., Smith, N.A.: A discriminative graph-based parser for the abstract meaning representation. In: Annual Meeting of the Association for Computational Linguistics (2014)
18. Fülöp, Z., Kühnemann, A., Vogler, H.: A bottom-up characterization of deterministic top-down tree transducers with regular look-ahead. Inf. Process. Lett. **91**, 57–67 (2004)
19. Fülöp, Z., Kühnemann, A., Vogler, H.: Linear deterministic multi bottom-up tree transducers. Theor. Comput. Sci. **347**, 276–287 (2005)

20. Ganian, R., Hliněný, P., Kneis, J., Langer, A., Obdržálek, J., Rossmanith, P.: Digraph width measures in parameterized algorithmics. Discrete Appl. Math. **168**, 88–107 (2014)
21. de Groote, P.: Towards a Montagovian account of dynamics. In: Semantics and Linguistic Theory 16 (2006)
22. Heim, I.: The semantics of definite and indefinite noun phrases. Ph.D. thesis, University of Massachusetts, Amherst (1982)
23. Hobbs, J.R., Shieber, S.M.: An algorithm for generating quantifier scopings. Comput. Linguist. **13**, 47–63 (1987)
24. Horrocks, I., Patel-Schneider, P.F., van Harmelen, F.: From SHIQ and RDF to OWL: the making of a web ontology language. Web Semant. Sci. Serv. Agents World Wide Web **1**, 7–26 (2003)
25. Kamp, H.: A theory of truth and semantic representation. In: Groenendijk, G., Janssen, T., Stokhof, M. (eds.) Formal Methods in the Study of Language. Mathematisch Centrum, Amsterdam (1981)
26. Keenan, E.L.: Further beyond the Frege boundary. In: van der Does, J., van Eijck, J. (eds.) Quantifiers, Logic, and Language. CSLI Publications, Amsterdam (1996)
27. Knight, K., et al.: Abstract meaning representation (AMR) annotation release 1.0 (2014). https://catalog.ldc.upenn.edu/LDC2014T12, http://amr.isi.edu/download.html
28. Kratzer, A.: Scope or pseudoscope? Are there wide-scope indefinites? In: Rothstein, S. (ed.) Events and Grammar. Springer, Dordrecht (1998)
29. Landau, I.: Control in Generative Grammar: A Research Companion. Cambridge University Press, New York (2013)
30. Landman, F.: Plurality. In: Lappin, S. (ed.) Handbook of Contemporary Semantic Theory, pp. 425–457. Oxford University Press, Oxford (1996)
31. Langekilde, I., Knight, K.: Generation that exploits corpus-based statistical knowledge. In: 36th Annual Meeting of the ACL, pp. 704–710 (1998)
32. Lebedeva, E.: Expression de la dynamique du discours a l'aide de continuations. Ph.D. thesis, Université de Lorraine (2012)
33. Maletti, A.: The power of weighted regularity-preserving multi bottom-up tree transducers. Int. J. Found. Comput. Sci. **26**(7), 293–305 (2015)
34. Montague, R.: Pragmatics. In: Thomason, R.H. (ed.) Formal Philosophy: Selected papers of Richard Montague. Yale University Press, New Haven, 1968/1974
35. Morgenstern, L., Davis, E., Ortiz, C.L.: Planning, executing, and evaluating the Winograd schema challenge. AI Mag. **37**(1), 50–54 (2016)
36. Muskens, R.: Order-independence and underspecification. In: Kamp, H., Partee, B. (eds.) Context-dependence in the Analysis of Linguistic Meaning, pp. 239–254. Elsevier (2004)
37. Partee, B.H., Borschev, V.: Genitives, relational nouns, and argument-modifier ambiguity. In: Lang, E., Maienborn, C., Fabricius-Hansen, C. (eds.) Modifying Adjuncts, pp. 67–112. Mouton de Gruyter, Berlin (2003)
38. Peters, P.S., Westerståhl, D.: Quantifiers in Language and Logic. Oxford University Press, Oxford (2006)
39. Szabolcsi, A.: Quantification. Cambridge University Press, New York (2010)
40. van Lambalgen, M., Hamm, F.: The Proper Treatment of Events. Blackwell, Oxford (2005)
41. Weischedel, R., Pradhan, S., Ramshaw, L., Kaufman, J., Franchini, M., El-Bachouti, M.: https://catalog.ldc.upenn.edu/docs/LDC2013T19/OntoNotes-Release-5.0.pdf (2015)

The Logic of Ambiguity: The Propositional Case

Christian Wurm[(✉)]

University of Düsseldorf, Düsseldorf, Germany
cwurm@phil.hhu.de

Abstract. We present a logical calculus extending the classical propositional calculus with an additional connective which has some features of substructural logic. This results in a logic which seems to be suitable for reasoning with ambiguity. We use a Gentzen style proof theory based on multi-contexts, which allow us to have two ways to combine formulas to sequences. These multi-contexts in turn allow to embed both features of classical logic as well as substructural logic, depending on connectives, which would be impossible with simple sequents. Finally, we present an algebraic semantics and a completeness theorem.

1 Introduction

The term LINGUISTIC AMBIGUITY designates cases where expressions of natural language give rise to two or more sharply distinguished meanings.[1] We let the ambiguity between two meanings m_1, m_2 be denoted by $m_1 \| m_2$. Ambiguity is usually considered to be a very heterogeneous phenomenon, and this is certainly true as far as it can arise from many different sources: from the lexicon, from syntactic derivations, semantic sources as quantifiers (this is sometimes reduced to syntax), and finally from literal versus collocational meanings. Despite this, we have recently argued (see [14]) that the best solution is to treat ambiguity consistently as part of semantics, because there are some properties which are consistently present irregardless of its source. The advantage of this unified treatment is that having all ambiguity in semantics, we can use all resources in order to resolve it and draw inferences from it (we will be more explicit below). Ambiguity is a really pervasive phenomenon in natural language, but mostly does not seem to pose any problems for speakers: in some cases, we do not even notice ambiguity, whereas in other cases, we can also perfectly reason with and draw inferences from ambiguous information:

(1) The first thing that strikes a stranger in New York is a big car.

Here for example, without any explicit reasoning, the conclusion that in New York there is at least one big car (and probably many more) seems sound to us. Hence we can easily draw inferences from ambiguous statements. This entails two things for us:

[1] This roughly distinguishes ambiguity from cases of vagueness [10].

A. Foret et al. (Eds.): FG 2017, LNCS 10686, pp. 88–104, 2018.
https://doi.org/10.1007/978-3-662-56343-4_6

1. We should rather not disambiguate *before* we start constructing semantics, as otherwise at least one reading remains *unavailable*, and soundness of inferences cannot be verified.
2. Hence we construct something as "ambiguous meanings", and it is perfectly possible to reason with them.

As regards 1., we have to add that disambiguating before interpreting is often plausible from a psychological point of view, and in many cases ambiguity can go completely unnoticed. However, from a logical point of view this prevents sound reasoning, and our goal here is to provide a theory for sound (and complete) reasoning with ambiguity, not a psychological theory.

We will present a logic of ambiguity which extends classical logic with an additional connective $\|$, a fusion style connective which is non-commutative (for now), non-monotonic in both directions, but which allows for both contraction and expansion. Moreover, it is self-dual, which is probably its most remarkable feature, a feature we find for example in classical bilinear logic (see [6]). Another remarkable property of our calculus is that we extend classical contexts to multi-contexts, hence embed classical logic into a larger logic. We find this idea briefly mentioned in [12], but to our knowledge this seems to be the first place where it is explicitly spelled out (though the idea of using multi-contexts is not new, see [2,5]).

The paper is structured as follows: we firstly introduce the key properties of ambiguity, in particular in relation to logic. Then we present the logic AL, an algebraic semantics, and establish its soundness and completeness.

2 Logic and Ambiguity

2.1 Background and Motivation

From a philosophical point of view, one often considers ambiguity to be a kind of "nemesis" of logical reasoning; for Frege for example, the main reason to introduce his logical calculus was that it was in fact unambiguous, contrary to natural language (but the discussion about the detrimental effect of ambiguity in philosophy can be traced back even to the ancient world, see [13], and is still going on, see [1]. On the other hand, in natural language semantics, there is a long tradition of dealing both with ambiguity and logic, since if we translate a natural language utterance into an unambiguous formal language such as predicate logic, ambiguity does not go away, but becomes visible by the fact that there are several translations. To consider a famous example:[2]

(2) `Every boy loves a movie.`

(3) $\exists x.\forall y.movie(x) \wedge (boy(y) \rightarrow loves(y,x))$

(4) $\forall y.\exists x.movie(x) \wedge (boy(y) \rightarrow loves(y,x))$

[2] Technically, this translation presupposes the existence of a boy, this however is irrelevant to our argument.

So we cannot simply translate natural language into logical representations (predicate logic or other), as there is no way to represent ambiguity in these languages. Note, by the way, that if we would use disjunction of (3) and (4), the formula would be logically equivalent to (3), hence this is no viable option. The standard way around the lack of functional interpretation is that we do not interpret *strings*, but rather *derivations*: one string has several syntactic derivations, and derivations in turn are functionally mapped to semantic representations (e.g. see [8]). The problem with this approach is that we basically ban ambiguity from semantics: we first make an (informed or arbitrary) choice, and then we construct an unambiguous semantics. Now this is a problem, as we have seen above:

1. If we simply pick one reading, we cannot know whether a conclusion is generally valid or not, because we necessarily discard some information.
2. To decide on a reading, we usually use semantic information; but if we choose a reading before constructing a semantic representation, how are we suppose to decide?

Now these two reasons indicate that we should not prevent ambiguity from entering semantics, because semantics is where we need it, and if it is only to get rid of it. But once ambiguity enters into semantics, we have to reason about its combinatorial, denotational and inferential properties.

For reasons of space, we will here only briefly expose what for us are the key features of ambiguity. For more extensive treatment, we refer the reader to [14]. Our exposition briefly lays out what are the challenges in developing a logic of ambiguity, and what are the key features it should have. We also want to quickly address the main reasons why ambiguity cannot be adequately treated with disjunction, which is a long-lasting misunderstanding among many scholars, even though this has been recognized many years ago, see for example [11].

2.2 Key Aspects of Ambiguity

1. Universal Distribution. For the combinatorics of $\|$, the most prominent (though only recently focussed, see [14]) feature of ambiguity is the fact that it equally distributes over all other connectives. To see this, consider the following examples:

(5) a. `There is a bank.`
    b. `There is no bank.`

(5-a) is ambiguous between m_1 = "there is a financial institute" and m_2 = "there is a strip of land along a river". When we negate this, the ambiguity remains, with the negated content: (5-b) is ambiguous between n_1 = "there is no financial institute" and n_2 = "there is no strip of land along a river", and importantly, the relation between the two meanings n_1 and n_2 is intuitively exactly the same as the one between m_1 and m_2. This distinguishes an ambiguous expression such as `bank` from a hypernym as `vehicle`, which is just more *general* than the meanings "car" and "bike":

(6) a. `There was a vehicle.`
b. `There was no vehicle.`

(6-a) means: "there was a car *or* there was a bike *or*..."; but (6-b) rather means: "there was no car *and* there was no bike *and*...". Hence when negated, the relation between the meanings changes from a disjunction to a conjunction (as we expect from a classical logical point of view); but for ambiguity, nothing like this happens: the relation remains invariant. This also holds for all other logical operations (see [14]). This invariance is the first point where we see a clear difference between ambiguity and disjunction. This property of *universal distribution* seems to be strongly related to another observation: we can treat ambiguity as something which happens in semantics (as we do here), or we can treat it as a "syntactic" phenomenon, where "syntactic" is to be conceived in a very broad sense. In our example, this would be to say: there is not one word (as form-meaning pair) `bank`, but rather two words $\texttt{bank}_1$ and $\texttt{bank}_2$, bearing different meanings. The same holds for genuine syntactic ambiguity: one does not assume that the sentence `I have seen a man with a telescope` has strictly speaking two meanings, one rather assumes it has two derivations (thus the string represents really two distinct sentences), where each derivation comes with a single meaning. Universal distribution is what makes sure that semantic and syntactic treatment are completely parallel: every operation f on an ambiguous meaning $m_1\|m_2$ equals an ambiguity between two (identical) operations on two distinct meanings, hence

(7) $$f(m_1\|m_2) = f(m_1)\|f(m_2)$$

Note that in cases where we combine ambiguous meanings with ambiguous meanings, this leads to an exponential growth of ambiguity, as is expected. Hence universal distribution is what creates the parallelism between semantic and syntactic treatment of ambiguity. This means: strictly speaking, we do not even need to argue whether ambiguity is a syntactic or semantic phenomenon – because the result in the end should be the same, it is of no relevance where ambiguity comes from. However, as soon as we start to *reason* with ambiguity, a unified semantic treatment will only have advantages, as all information is in one place. As we only consider propositional logic, (7) reduces to

(8) $$\neg(\alpha\|\beta) \equiv \neg\alpha\|\neg\beta$$
(9) $$(\alpha\|\beta) \vee \gamma) \equiv (\alpha \vee \gamma)\|(\beta \vee \gamma)$$
(10) $$(\alpha\|\beta) \wedge \gamma) \equiv (\alpha \wedge \gamma)\|(\beta \wedge \gamma)$$

By convention, we use symbols as m_1, m_2 if we speak about (propositional) linguistic meanings, symbols like a, b, c when we speak about arbitrary algebraic objects; Greek letters α, β etc. will be reserved for logical formulas. Logically speaking, this means that $\|$ is SELF-DUAL: $\|$ preserves over negative contexts such as negation, as fusion in [6] (this logic is however used for a very different purpose, namely syntactic analysis).

Entailments. An ambiguity $m_1\|m_2$ is generally characterized by the fact the speaker intends one of m_1 or m_2. The point is: we do not know which one of the two, as for example in

(11) Give me the dough!

From this simple fact, we can already deduce that for arbitrary formulas $\phi, \alpha, \beta, \chi$ in the logic of ambiguity, if $\phi \vdash \alpha \vdash \chi$ and $\phi \vdash \beta \vdash \chi$ hold, then $\phi \vdash \alpha\|\beta \vdash \chi$ holds, hence in particular, $\alpha \wedge \beta \vdash \alpha\|\beta \vdash \alpha \vee \beta$. But: we cannot reduce $\alpha\|\beta$ to neither α nor β: we have $\alpha \not\vdash \alpha\|\beta$ and $\beta \not\vdash \alpha\|\beta$, and also $\alpha\|\beta \not\vdash \alpha$ and $\alpha\|\beta \not\vdash \beta$. This is because our logic is supposed to model which inferences are valid in *every* case (i.e. under every intention), not in *some* cases, and all the latter entailments are all invalid in some cases. Considering (11), the speaker either means "pastry" or "money", but he might complain either when you given him the money or when you give him the pastry. Hence $\|$ does not coincide with any classical connective and is not Boolean definable. It is actually a **substructural** connective (see [12] for an introduction), behaving similar as fusion in linear logic: in particular, it does not allow for weakening (we will make this precise below). Note that this also illustrates how ambiguity behaves differently from disjunction.

Conservative Extension. In particular in connection with logic, it should be clear that our logical calculus of ambiguity should be a conservative extension of the classical calculus. The reason is that even if we include ambiguous propositions, unambiguous propositions should behave as they used to before – if there are new entailments, they should only concern ambiguous propositions.

Monotonicity/Consistency. Imagine someone telling you something about `banks`, and as he goes on, you discover that what he says does not make any sense to you. In the end, you notice that he has been using the term `bank` with different meanings in different utterances. At this point, you will obviously have to consider his entire discourse meaningless: how can you possibly reconstruct what meaning was intended in which utterance? Hence reasoning with ambiguous information presupposes UNIFORM USAGE: terms with several senses must be used consistently in one sense. And in fact, arguments with ambiguous terms fail if this principle of uniform usage is violated; this marks the line between their *use* and *abuse*. Hence we have the following principle:

(UU) In a given context, an ambiguous statement must be used consistently in *only one* sense.

This is of course very arguable, not only because the notion of "context" remains vague, but also because we can use the same word with different meanings in the same sentence, as in `I spring over a spring in spring.`[3] There is a lot to say on this issue, for us however (UU) remains a technical necessity.

[3] Thanks to an anonymous reviewer for this example!.

(UU) also clarifies the following important point: whereas in classical reasoning, we have inconsistency by logical contradiction, in reasoning with ambiguity, there is another source of inconsistency, namely inconsistent usage of ambiguous terms. Logically, uniform usage has its counterpart in the following inference (we denote this by monotonicity):

$$\text{(monotonicity)}\ \frac{\alpha \vdash \gamma \quad \beta \vdash \delta}{\alpha\|\beta \vdash \gamma\|\delta}$$

In our logic, this will rule will be admissible, though we formulate rules in a more general way for technical reasons.

There are some more important properties of ambiguity, such as the one that it is not productive and hence in natural language, we do not find arbitrary ambiguities. However, as neither there seems to be an apriori restriction on ambiguity, we will ignore this issue (and some others) for the moment.

2.3 A Note on the Standard Treatment

As we have said, in natural language semantics it is already common to represent ambiguous meanings in one way or other. The standard approach for representing ambiguity (as e.g. in the quantifier case) is to use a sort of meta-semantics,[4] whose expressions underspecify the logical representations (see for example [3]). Assume our "unambiguous" language is the logic $\mathcal{L}$; and call the meta-language $\mathcal{M}$, where for example χ is a formula of $\mathcal{M}$ underspecifying the two formulas α, β of $\mathcal{L}$ (for example (3) and (4)). But now that we have this meta-language $\mathcal{M}$ of our logic $\mathcal{L}$, there are new questions:

1. How do we interpret terms of $\mathcal{M}$?
2. How do we provide connectives of $\mathcal{M}$ with a compositional semantics?
3. What are the inferences both in $\mathcal{L}$ and $\mathcal{M}$ we can draw from terms in $\mathcal{M}$?

Once we start seriously addressing these questions, we see that moving to a meta-language does not solve any problems – at best, it removes them from our sight. We usually do have a compositional semantics and consequence relation for $\mathcal{L}$; for $\mathcal{M}$ we do not. Hence $\mathcal{M}$ fails to have the most basic features of a semantics, unless, of course, $\mathcal{M}$ itself is a logic with consequence relation and compositional semantics. But in this case, considering that $\mathcal{M}$ should conservatively extend $\mathcal{L}$, it seems to be much more reasonable to include the new operator for ambiguity into our object language $\mathcal{L}$. And this is exactly what we do here. From this example it becomes once more clear that ambiguity cannot be reasonably interpreted the same way as disjunction: because $\mathcal{L}$ in any normal case already has disjunction, and there would be no need at all for $\mathcal{M}$.

---

[4] Actually, this would be a meta-metalanguage, because logical representations are already a form of representation of real meanings.

3 The Ambiguity Logic AL

3.1 Multi-sequents and Contexts

We want to advert the reader that from the presentation of AL, it will not be immediately clear how it relates to ambiguity. This will be much more obvious for its semantics, universal distribution algebras, which we present in Sect. 4; hence if the reader is interested in the motivation rather than the logic, we advise him to first consider Sect. 4. The logic AL is a conservative extension of classical (propositional) logic, that is, it derives all and only the valid sequents of classical logic in the language of the latter, but it has an additional connective $\|$, with which we can derive additional valid sequents. $\|$ is not very exotic from the point of view of substructural logic: it is a fusion-style operator, which allows for contraction and expansion (its inverse), but not for weakening; we can present it both in a commutative and non-commutative version. Our approach differs from the usual approach to substructural logic in that we *extend* classical logic with a substructural connective, whereas usually, one considers logics which are proper fragments of classical logic. In order to make this possible, we have to go beyond the normal sequent calculus: we still have sequents, but we have different types of contexts: one we denote by $\natural(...)$, which basically embeds classical logic, one we denote by $\Diamond(...)$, which allows to introduce the new connective $\|$. These contexts thus differ in what kind of connectives we can introduce in them, and what kind of structural rules are allowed in them. For technical reasons, there will also be a third, negative context $\flat(...)$, which has however a less "deep" meaning. The first two contexts can be arbitrarily embedded within each other, whereas the negative context is restricted to single formulas. We refer to the symbols $\natural, \Diamond, \flat$ as **modalities** (but they do not really relate to modal logic).

We call the resulting structures **multi-contexts**, a pair $\Delta \vdash \Gamma$, with Δ, Γ multi-contexts we call a **multi-sequent**, and the calculus a **multi-sequent calculus**. We have found this idea briefly mentioned as a way to approach substructural logic in [12], and structures similar to multi-contexts are found in [2]. Our approach is particular in that we actually extend classical propositional contexts, and as AL is but one particular instance of multi-sequent logics, we think that this field definitely deserves further study.

In order to increase readability, we distinguish contexts both by the symbols $\natural, \Diamond, \flat$, and by the type of period we use to separate formulas/contexts. This will be ',' in the classical context, so $\natural(\alpha, \beta)$ is a well-formed (classical) context. Here ',' corresponds to $\wedge$ left of $\vdash$, and to $\vee$ right of $\vdash$, and allows for all structural rules. In the ambiguous context, we use ';', hence $\Diamond(\alpha; \beta)$ is a well-formed (ambiguous) context. ';' corresponds to $\|$, is self-dual, and allows for some structural rules such as contraction, but not for others, such as weakening or commutativity. For the negative context $\flat$, this problem will not arise, as it is strictly unary. Formulas are defined as usual, we have a set *Var* of propositional variables, and:

- if $p \in Var$, then $p \in \texttt{WFF}$;
- if $\phi, \chi \in \texttt{WFF}$, then $\phi \wedge \chi, \phi \vee \chi, \phi \| \chi, \neg\phi \in \texttt{WFF}$;
- nothing else is in WFF.

Next, we define multi-contexts; for sake of brevity, we refer to them simply as **contexts**.

1. $\natural(\epsilon)$, where ϵ is the empty sequence, is a well-formed, classical context, which we also call the empty context.
2. If $\gamma \in \texttt{WFF}$,then $\natural(\gamma)$ is a well-formed, classical context.
3. If $\gamma \in \texttt{WFF}$, then $\flat(\gamma)$ is a well-formed, negative context.
4. If $\Gamma_1, ..., \Gamma_i$ are well-formed contexts, then $\natural(\Gamma_1, ..., \Gamma_i)$ is a well-formed, classical context.
5. If Γ_1, Γ_2 are well-formed, non-empty contexts, then $\Diamond(\Gamma_1; \Gamma_2)$ is a well formed ambiguous context.

Note that $\flat$ is a strictly unary modality, whereas $\Diamond$ is strictly binary. This choice is somewhat arbitrary, but seems to be the most elegant way to prevent some technical problems. $\natural$ has no restriction in this sense. $\Gamma \vdash \Delta$ is a **well-formed multi-sequent**, if both Γ, Δ are well-formed, *classical* contexts. The calculus with all modalities is somewhat clumsy to write, so we have a number of **conventions** for increasing readability:

- We generally omit unary classical contexts; hence $\Diamond(\alpha; \beta)$ is short for $\Diamond(\natural(\alpha); \natural(\beta))$. We never omit negative contexts, which are always unary.
- In the same vein, we omit the outermost context in multi-sequents. We can do this because it always is $\natural(...)$, otherwise the sequent would not be well-formed. As a special case, we omit the empty context $\natural()$. Hence $\vdash \alpha$ is a shorthand for $\natural() \vdash \natural(\alpha)$ etc.
- We write Γ to refer to arbitrary contexts, so α, Γ is a shorthand for $\natural(\natural(\alpha), \Gamma)$;
- We write $\Gamma[\alpha]$ to refer to a subformula α of a context Γ; same for $\Gamma[\Delta]$, where Δ is a sub-context.
- We write $\Gamma[_\natural\alpha]$ etc. in order to indicate that α does occur in the scope of $\natural$, that is, the smallest sub-context containing it is classical.

We urge the reader to be careful: we will make full use of these conventions already in the presentation of the sequent calculus. The reason is that only this way, it will be plain obvious that our calculus is a nice extension of classical logic. Moreover, we aim to formulate the calculus in a way to make the structural rules of contraction and weakening admissible, as far as they are desired (see [9] for background), though we cannot prove these properties here for reasons of space. For the same reason, we skip the proof of basic properties such as the fact that all rules preserve well-formedness of multi-sequents, which in fact is not entirely trivial.

3.2 The Classical Context and Its Rules

The modality $\natural$ (partly) embeds the classical calculus; hence we have the following well-known rules:

(ax) $\overline{\alpha, \Gamma \vdash \alpha, \Delta}$

$$(\wedge\mathrm{I})\ \frac{\Gamma[_\natural\alpha,\beta]\vdash\Theta}{\Gamma[_\natural\alpha\wedge\beta]\vdash\Theta}\qquad(\mathrm{I}\wedge)\ \frac{\Gamma\vdash\Theta[\alpha]\quad\Gamma\vdash\Theta[\beta]}{\Gamma\vdash\Theta[\alpha\wedge\beta]}$$

$$(\vee\mathrm{I})\ \frac{\Gamma[\alpha]\vdash\Theta\quad\Gamma[\beta]\vdash\Theta}{\Gamma[\alpha\vee\beta]\vdash\Theta}\qquad(\mathrm{I}\vee)\ \frac{\Gamma\vdash\Theta[_\natural\alpha,\beta]}{\Gamma\vdash\Theta[_\natural\alpha\vee\beta]}$$

Note that in (I$\wedge$), ($\vee$I) there is no requirements regarding the context. ($\wedge$I) and (I$\vee$) show how $\wedge,\vee$ correspond to ',', depending on the side of $\vdash$. The classical rules of negation introduction are *not* part of the calculus, however, they are admissible in it; this will be clear when we introduce the negative context $\flat$. In the following, we have the three structural rules of classical logic; these rules are of course bound to the classical context. We conjecture that weakening and contraction are admissible in the calculus (usual argument of reducing the degree of the rule), so the only rule we really need is commutativity.

$$(\natural\mathrm{comm})\ \frac{\Gamma[_\natural\Psi,\Theta]}{\Gamma[_\natural\Theta,\Psi]}.\qquad(\natural\mathrm{weak})\ \frac{\Gamma[_\natural\Delta]}{\Gamma[_\natural\Delta,\Psi]}\qquad(\natural\mathrm{contr})\ \frac{\Gamma[_\natural\Delta,\Delta]}{\Gamma[_\natural\Delta]}$$

This notation means that the rules can be equally applied on both sides of $\vdash$. Note that we have all these rules not for formulas, but for contexts (recall that in our notation, a formula is just a shorthand for an atomic context anyway). Finally, note that we cannot explicitly introduce $\natural$ at any point, and neither eliminate it explicitly. But our rules have a number of implicit eliminations of $\natural$, for example when we combine two formulas to one.

3.3 The Ambiguous Context and Its Rules

$\Diamond$ is a strictly binary modality, and hence there should be no way to introduce single formulas in this context. The introduction rules for $\Diamond$ are as follows:

$$(\Diamond\mathrm{I1})\ \frac{\Gamma,\Lambda\vdash\Delta,\Psi\quad\Theta,\Lambda\vdash\Phi,\Psi}{\Diamond(\Gamma;\Theta),\Lambda\vdash\Diamond(\Delta;\Phi),\Psi}\quad(\Diamond\mathrm{I2})\ \frac{\Gamma,\Lambda\vdash\Delta\quad\Theta,\Lambda\vdash\Delta}{\Diamond(\Gamma;\Theta),\Lambda\vdash\Delta}\quad(\Diamond\mathrm{I3})\ \frac{\Gamma\vdash\Delta,\Lambda\quad\Gamma\vdash\Phi,\Lambda}{\Gamma\vdash\Diamond(\Delta;\Phi),\Lambda}$$

Note that if we would allow the empty context in $\Diamond(;)$, then the rules ($\Diamond$I2),($\Diamond$I3) are just particular instances of ($\Diamond$I1). Hence all these rules can be seen as special instances of a single one, which would however be tedious to write down. Alternatively, we can derive ($\Diamond$I2), ($\Diamond$I3) from ($\Diamond$I1) with ($\Diamond$contr). However, we rather want ($\Diamond$contr) to be admissible, as it causes an infinite search space. Consider also the particular instance of ($\Diamond$I1) where Λ,Ψ are empty: here we can see that these rules are in a sense a generalization of $\bullet$-introduction in the Lambek-calculus, and simply generalize (monotonicity) we mentioned above. There are two (parallel) introduction rules for $\|$:

$$(\|\mathrm{I})\ \frac{\Gamma[\Diamond(\alpha;\beta)]\vdash\Theta}{\Gamma[\alpha\|\beta]\vdash\Theta}\qquad(\mathrm{I}\|)\ \frac{\Gamma\vdash\Theta[\Diamond(\alpha;\beta)]}{\Gamma\vdash\Theta[\alpha\|\beta]}$$

At the same time, these rules eliminate the $\Diamond$-context. There are two structural rules in $\Diamond$-context, namely associativity and contraction (we do for now not allow commutativity), where for the latter we conjecture admissibility.

$$(\Diamond\text{ass})\ \frac{\Psi[\Diamond(\Gamma;\Diamond(\Delta;\Theta))]}{\Psi[\Diamond(\Diamond(\Gamma;\Delta);\Theta))]} \qquad (\Diamond\text{contr})\ \frac{\Gamma[\Diamond(\alpha;\alpha)]}{\Gamma[\Diamond(\alpha)]}$$

Here double lines indicate that the rule works in both directions, and absence of $\vdash$ means rules work equally on both sides. We still need a rule which ensures that we will always satisfy the distributive laws as required. The introduction rules for $\Diamond$ are already sufficient to derive $\wedge$, $\vee$-distribution over $\|$ *in the atomic case*; however, they interact with the negation rules (see below) in a manner not sufficiently strong to ensure that this holds for $\dashv\vdash$ as a *congruence* as long as we do not use cut. Therefore, we include the following unproblematic rule:

$$(\text{distr})\ \frac{\Gamma[\natural\Diamond(\Delta;\Psi),\Theta_1] \quad \Gamma[\natural\Diamond(\Delta;\Psi),\Theta_2]}{\Gamma[\Diamond(\natural(\Delta,\Theta_1);\natural(\Psi,\Theta_2))]}$$

Θ_1, Θ_2 are distinct, as otherwise, we have a form of expansion, which is problematic for cut elimination. This rule slightly generalize normal distribution: if $\Theta_1 = \Theta_2$, we get simple distribution, and in this special case, the rule is also *invertible*, that is, its inversion is admissible in the calculus. It is not difficult to show that without (distr), we cannot eliminate cut.

3.4 The Negative Context

$\flat$ marks the negative context. We design it in a way such that it subsumes classical negation rules. For this reason, we have to make sure no (classical) structural rules are applied in this context: in particular, using weakening in $\flat()$ would lead us into trouble, as the meaning of ',' changes with position with respect to $\vdash$. This is why $\flat$ only applies to formulas. We need this context to derive the distributional laws for $\|$ and negation, which are not derivable so far.

$$(\flat\text{I})\ \frac{\Gamma[\natural(\alpha)]}{\Gamma[\flat(\neg\alpha)]} \qquad (\flat\text{distr})\ \frac{\Gamma[\Diamond(\flat(\phi);\flat(\chi))]}{\Gamma[\flat(\phi\|\chi)]}$$

Again, these rules operate equally on both sides of $\vdash$. Note that this is the only occasion where we explicitly write $\natural(\phi)$, because in this case, the classical modality is actually cancelled and replaced by $\flat$. Note also that in ($\flat$distr), we introduce $\|$ and delete an ambiguous context.

$$(\flat\text{E})\ \frac{\Gamma,\flat(\alpha)\vdash\Delta}{\Gamma\vdash\Delta,\alpha} \qquad (\text{E}\flat)\ \frac{\Gamma\vdash\Delta,\flat(\alpha)}{\Gamma,\alpha\vdash\Delta}$$

Hence the modality is simply eliminated by changing the position. It is easy to see that this subsumes classical negation rules: it splits one step into two, thereby allowing for negation distribution over $\|$ as an intermediate step. The best way to think of $\flat$ is maybe to assume that every formula has an atomic polarity attached, which is either positive in the case of $\natural$, or negative in the case of $\flat$.

3.5 Cut Rules

We now present the cut rule. Its adaption to our multi-sequents is not entirely straightforward, as we have to be sensitive to different contexts: cut needs to be aware of the modality of the cut formula (the formula being substituted), because otherwise we might insert positive contexts into negative contexts, which would be unsound. As $\Diamond$ is a strictly binary modality, it does not play a role for cut, as it is never the modality attached to a cut formula.

$$(\natural\text{cut})\ \frac{\Gamma[\natural(\alpha)] \vdash \Psi \quad \Delta \vdash \natural(\alpha), \Theta}{\Gamma[\Delta] \vdash \Psi, \Theta} \qquad (\Diamond\text{cut})\ \frac{\Gamma[\flat(\alpha)] \vdash \Psi \quad \Delta \vdash \flat(\alpha), \Theta}{\Gamma[\Delta] \vdash \Psi, \Theta}$$

The two rules could be obviously merged into one, if we used a meta-variable for $\natural, \flat$; this would however not really simplify things. These cut rules ensure transitivity and congruence without any special cases to consider. Importantly, as every context has a particular modality, also the sequence inserted by cut comes with a modality – but it need not be the same as the one of the cut-formula!

We define the notion of a derivation as usual as a proof-tree with the leaves being the instances of (ax); a multi-sequent $\Gamma \vdash \Delta$ is derivable if it is the root of such a proof-tree. In this case, we write $\Vdash_{\mathsf{AL}} \Gamma \vdash \Delta$, meaning the sequent is derivable in AL.

4 Semantics of AL

4.1 Universal Distribution Algebras

We now introduce a class of algebraic models for AL. For reasons of space, we cannot dwell on algebraic properties of this class, though they are also quite useful for understanding AL. We call this class **universal distribution algebras** or **UDA**. From the axioms, it will be easy to see that it is also a nice model for ambiguity. A universal distribution algebra is an algebra $\mathbf{U} = (U, \wedge, \vee, \sim, \|, 0, 1)$, where $(U, \wedge, \vee, \sim, 0, 1)$ is a **Boolean algebra** (for background on Boolean algebras see [4,7]), and $\|$ is a binary function satisfying the following axioms ($a \leq b$ is an abbreviation for $a \wedge b = a$ or equivalently $a \vee b = b$):

$$\begin{aligned}
&(\|1) && (a\|b) \wedge c = (a \wedge c)\|(b \wedge c)\\
&(\|2) && \sim(a\|b) = \sim a\|\sim b\\
&(\text{ass}) && (a\|b)\|c = a\|(b\|c)\\
&(\text{inf}) && a \wedge b \leq a\|b \leq a \vee b\\
&(\text{mon}) && a\|b \leq (a \vee c)\|(b \vee d)
\end{aligned}$$

($\|1$) and ($\|2$) make sure $\|$ has the property of universal distribution ($\vee$ is redundant). (ass) is clear; (inf) regulates the relation '$\leq$' between ambiguous and non-ambiguous objects; (mon) the relation '$\leq$' between ambiguous objects. Note that (inf) is partly redundant, as $a\|b \leq a \vee b$ entails $a \wedge b \leq a \| b$ and vice versa (use complementation). Spelling out (mon), we can see it is basically a sort of distributive law. It is also easy to see that (mon) is equivalent to monotonicity: it states that if $a \leq a'$, $b \leq b'$, then $a\|b \leq a'\|b'$. The formulation we chose shows that **UDA** is a variety. Note that $\|$ does not have a unit element, because intuitively, there is no unit for ambiguity. In presence of distributive laws, (inf) is equivalent to (id) $a\|a = a$: (id) entails (inf), because $(a\|b) \wedge (a \wedge b) = (a \wedge b)\|(a \wedge b) = a \wedge b$, and hence by definition of $\leq$, $a \wedge b \leq a\|b$; parallel for $a \vee b$. Conversely, (inf) entails (id), because then $a = a \wedge a \leq a\|a \leq a \vee a = a$. In **UDA**, there are many interesting properties we cannot state here for reasons of space. However, from the axioms it is clear that **UDA** presents a nice model for ambiguity.

4.2 Interpretations of AL

The interpretation of AL into **UDA** is straightforward, but we have to spell it out nonetheless. We define interpretations for contexts; this is necessary for the usual inductive soundness proof. Assume $\mathbf{U} \in \mathbf{UDA}$, $\sigma : \mathit{Var} \to U$ is then an atomic interpretation. We define two interpretation functions $\overline{\sigma}, \underline{\sigma}$ by:

1. $\overline{\sigma}(p) = \sigma(p) = \underline{\sigma}(p)$, for $p \in \mathit{Var}$.
2. $\overline{\sigma}(\phi \wedge \chi) = \overline{\sigma}(\phi) \wedge \overline{\sigma}(\chi) = \underline{\sigma}(\phi \wedge \chi)$
3. $\overline{\sigma}(\phi \vee \chi) = \overline{\sigma}(\phi) \vee \overline{\sigma}(\chi) = \underline{\sigma}(\phi \vee \chi)$
4. $\overline{\sigma}(\neg\chi) = {\sim}\overline{\sigma}(\chi) = \underline{\sigma}(\neg\chi)$
5. $\overline{\sigma}(\phi\|\chi) = \overline{\sigma}(\phi)\|\overline{\sigma}(\chi) = \underline{\sigma}(\phi\|\chi)$
6. $\overline{\sigma}(\natural(\Gamma_1, ..., \Gamma_i)) = \overline{\sigma}(\Gamma_1) \vee ... \vee \overline{\sigma}(\Gamma_i)$
7. $\underline{\sigma}(\natural(\Gamma_1, ..., \Gamma_i)) = \underline{\sigma}(\Gamma_1) \wedge ... \wedge \underline{\sigma}(\Gamma)$
8. $\overline{\sigma}(\flat(\phi)) = {\sim}(\overline{\sigma}(\phi)) = \underline{\sigma}(\flat(\phi))$
9. $\overline{\sigma}(\Diamond(\Gamma; \Delta)) = \overline{\sigma}(\Gamma)\|\overline{\sigma}(\Delta)$
10. $\underline{\sigma}(\Diamond(\Gamma; \Delta)) = \underline{\sigma}(\Gamma)\|\underline{\sigma}(\Delta)$

As is easy to see, $\overline{\sigma}$ and $\underline{\sigma}$ coincide on formulas, and hence in the formula case there is no reason to distinguish them. They also coincide in their interpretation of ';', but as there might be a classical context embedded, it is important to distinguish them. With $\flat$, there is no need to keep them distinct, as this context only embeds formulas.

We define truth in a model as usual: $\mathbf{U}, \sigma \models \Gamma \vdash \Delta$ iff $\underline{\sigma}(\Gamma) \leq_U \overline{\sigma}(\Delta)$; as a special case, we have $\mathbf{U}, \sigma \models \vdash \Delta$ iff $1_U \leq_U \overline{\sigma}(\Delta)$ and $\mathbf{U}, \sigma \models \Delta \vdash$ iff $\underline{\sigma}(\Delta) \leq_U 0_U$. Moreover, we define the notion of validity as usual by $\mathbf{UDA} \models \Gamma \vdash \Delta$ (stating that $\Gamma \vdash \Delta$ is valid) iff for all $\mathbf{U} \in \mathbf{UDA}$, $\sigma : \mathit{Var} \to U$, we have $\mathbf{U}, \sigma \models \Gamma \vdash \Delta$. We now prove soundness and completeness of **UDA**-semantics for AL, that is, $\mathbf{UDA} \models \Gamma \vdash \Delta$ iff $\Vdash_{\mathsf{AL}} \Gamma \vdash \Delta$. We start with a section on soundness.

4.3 Soundness for AL

In this section, we only prove the following lemma:

Lemma 1 *(Soundness). If* $\Vdash_{\mathsf{AL}} \Gamma \vdash \Delta$*, then* **UDA** $\models \Gamma \vdash \Delta$.

Proof. We make the usual induction over proof rules, showing they preserve correctness. We omit this for the classical rules for which the standard proofs can be taken over with minor modifications.

- ▶ ($\Diamond$I1) Assume $\Gamma, \Lambda \vdash \Delta, \Psi$ and $\Theta, \Lambda \vdash \Phi, \Psi$ are true in a model. Then by (mon) $\Diamond(\natural(\Gamma, \Lambda); \natural(\Theta, \Lambda)) \vdash \Diamond(\natural(\Delta, \Psi); \natural(\Phi, \Psi))$ is true, too. It is now easy to check that by distributive laws,

$$\overline{\sigma}(\Diamond(\natural(\Gamma, \Lambda); \natural(\Theta, \Lambda))) = \overline{\sigma}(\natural(\Diamond(\Gamma; \Theta), \Lambda))$$
$$\overline{\sigma}(\Diamond(\natural(\Delta, \Psi); \natural(\Phi, \Psi))) = \overline{\sigma}(\Diamond(\Delta; \Phi), \Psi)$$

  Same for $\underline{\sigma}$.
- ▶ ($\Diamond$I2),($\Diamond$I3) are just particular instances of ($\Diamond$I1), provided we use ($\Diamond$contr), which is sound by idempotence (which in turn is equivalent to (inf)).
- ▶ ($\|$I),(I$\|$),(ass): the former are sound, because antecedent and consequent have actually the same interpretation; the latter is obvious.
- ▶ ($\flat$I) is sound because the law of double negation holds in **UDA**: hence $\overline{\sigma}(\flat(\neg\phi)) = \overline{\sigma}(\phi)$, same for $\underline{\sigma}$, hence the claim follows easily.
- ▶ ($\flat$distr) is sound because of ($\|$2), negation distribution.
- ▶ ($\flat$E),(E$\flat$) As the function $\sim$ is a bijection in all Boolean algebras, soundness of these rules (eliminating a $\sim$) is equivalent to the soundness of the classical negation introduction rules (technically, it is their contraposition).
- ▶ ($\vee$I) As contexts on the left of $\vdash$ are interpreted as terms over $\|, \wedge, \sim$, we show that

$$(\#) \qquad a\|(b \vee b')\|c \leq (a\|b\|c) \vee (a\|b'\|c)$$
$$(+) \qquad a \wedge (b \vee b') \wedge c \leq (a \wedge b \wedge c) \vee (a \wedge b' \wedge c)$$
$$(*) \qquad {\sim}(a \vee b) \leq {\sim}a \vee {\sim}b$$

  from which the soundness of the rule follows by an easy induction on the complexity of the context. (+) and (*) are obvious and well-known; we prove

$$\begin{aligned}(a\|b\|c) \vee (a\|b'\|c) &= (a \vee (a\|b'\|c))\|(b \vee (a\|b'\|c))\|(c \vee (a\|b'\|c))\\ &\geq a\|(b \vee (a\|b'\|c))\|c \text{ (by (mon))}\\ &= a\|(b \vee a)\|(b \vee b')\|(b \vee c))\|c \text{ (by ($\|$1))}\\ &\geq a\|a\|(b \vee b')\|c\|c \text{ (by (mon))}\\ &\geq a\|(b \vee b')\|c \text{ (by (id))}\end{aligned}$$

- ▶ (I$\wedge$) A parallel argument to ($\vee$I): invert $\geq$ and $\leq$, and show that $(a\|b\|c) \wedge (a\|b'\|c) \leq a\|(b \wedge b')\|c$. Then we can perform the same induction on contexts.
- ▶ (distr) We just consider the case on the left of $\vdash$; the other case is parallel. So assume $\Gamma[(\Delta; \Psi), \Theta_1] \vdash \Xi$ and $\Gamma[(\Delta; \Psi), \Theta_2] \vdash \Xi$ are true in a model. Assume moreover that θ_1, θ_2 are formulas such that $\underline{\sigma}(\theta_1) = \underline{\sigma}(\Theta_1)$ and $\underline{\sigma}(\theta_2) = \underline{\sigma}(\Theta_2)$, which obviously exist. We can then see (because of soundness of $\vee$-rules) that $\Gamma[(\Delta; \Psi), \theta_1 \vee \theta_2] \vdash \Xi$ is true, and by distributive laws, $\Gamma[(\natural(\Delta, \theta_1 \vee \theta_2); \Psi, \natural(\theta_1 \vee \theta_2))] \vdash \Xi$ is also true. Now as $\underline{\sigma}(\Theta_1) = \underline{\sigma}(\theta_1) \leq \underline{\sigma}(\theta_1 \vee \theta_2)$, same for θ_2, it follows that $\Gamma[(\natural(\Delta, \Theta_1); \natural(\Psi, \Theta_2))] \vdash \Xi$ is also true. For the right side of $\vdash$, we just use $\wedge$ instead of $\vee$, $\overline{\sigma}$ instead of $\underline{\sigma}$.

- ($\natural$cut) We use the well-known fact that in Boolean algebras, we have $a \wedge \neg b \leq c$ iff $a \leq c \vee b$. Assume both $\Gamma[\natural\alpha] \vdash \Psi$ and $\Delta \vdash \natural\alpha, \Theta$ are true in a model, and let $\theta \in \texttt{WFF}$ be a formula such that $\overline{\sigma}(\theta) = \overline{\sigma}(\Theta)$. Then $\Delta, \neg\theta \vdash \natural\alpha$ is true, and by congruence, so is $\Gamma[\natural(\Delta, \neg\theta)] \vdash \Psi$. Now we make an intermediate step:
  $\underline{\sigma}(\Gamma[\Delta], \Theta) \leq \underline{\sigma}(\Gamma[\natural(\Delta, \Theta)])$. This can be shown by an easy induction over Γ, the crucial step being that $(a\|b) \wedge c = (a \wedge c)\|(b \wedge c) \leq (a\|(b \wedge c))$ etc. So $\Gamma[\Delta], \neg\theta \vdash \Psi$ remains true, and by double negation elimination, so is $\Gamma[\Delta] \vdash \Psi, \theta$, where θ can be again replaced by Θ.
- ($\flat$cut) can be, as far as semantics is concerned, be conceived of a special case of ($\natural$cut), where $\alpha = \neg\alpha'$. But this reduction obviously only works on the semantic side, syntactically, the rule has to be kept distinct. □

4.4 Completeness for AL

We now present a standard algebraic completeness proof for AL and **UDA** via the Lindenbaum algebra for AL, denoted by **Linda**. Its carrier set M is the set of AL-formulas modulo logical equivalence: we write $\alpha \dashv\vdash \beta$ iff $\Vdash_{\mathsf{AL}} \alpha \vdash \beta$, $\Vdash_{\mathsf{AL}} \beta \vdash \alpha$. This relation is symmetric by definition, reflexive and transitive (by cut). We put $\alpha_{\dashv\vdash} = \{\beta : \beta \dashv\vdash \alpha\}$, and $M = \{\alpha_{\dashv\vdash} : \alpha \in \texttt{WFF}\}$. The next step will be to show that $\dashv\vdash$, more than an equivalence relation, is a *congruence* over connectives.

Lemma 2. *Assume* $\alpha_1 \dashv\vdash \beta_1$, $\alpha_2 \dashv\vdash \beta_2$. *Then for* $\star \in \{\wedge, \vee, \|\}$, $\alpha_1 \star \alpha_2 \dashv\vdash \beta_1 \star \beta_2$, *and* $\neg\alpha_1 \dashv\vdash \neg\beta_1$.

Proof. By cases; for all classical connectives, just use standard proof; for $\|$, this is no less straightforward. □

Hence we can use the equivalence classes irrespective of representatives and define, for $m, n \in M$:

- $m \wedge n = (\alpha \wedge \beta)_{\dashv\vdash}$, where $\alpha \in m, \beta \in n$
- $m \vee n = (\alpha \vee \beta)_{\dashv\vdash}$, where $\alpha \in m, \beta \in n$
- $m\|n = (\alpha\|\beta)_{\dashv\vdash}$, where $\alpha \in m, \beta \in n$
- $\sim m = (\neg\alpha)_{\dashv\vdash}$, where $\alpha \in m$
- $1 = (p \vee \neg p)_{\dashv\vdash}$, where $p \in Var$
- $0 = (p \wedge \neg p)_{\dashv\vdash}$, where $p \in Var$

Since our calculus subsumes the classical propositional calculus, the algebra $(M, \wedge, \vee, \sim, 0, 1)$ is a Boolean algebra, where the relation $\leq$ coincides with $\vdash$ (modulo equivalence). We prove it is a universal distribution algebra:

Lemma 3. $(M, \wedge, \vee, \sim, \|, 0, 1)$ *is a universal distribution algebra.*

Proof. As $\vdash$ corresponds to $\leq$, $=$ corresponds to $\dashv\vdash$. Hence equalities fall into two subclaims, which we sometimes treat separately.

($\|1$) i. $(a\|b) \wedge c \leq (a \wedge c)\|(b \wedge c)$.

$$\dfrac{\dfrac{\dfrac{a,c\vdash a \quad b,c\vdash c}{\Diamond(a;b),c\vdash\Diamond(a;c)}\,(\Diamond I1) \quad \dfrac{\Diamond(a;b),c\vdash c \quad \Diamond(a;b),c\vdash c}{\Diamond(a;b),c\vdash\Diamond(c;c)}\,(\Diamond I3)}{\Diamond(a;b),c\vdash\Diamond(a\wedge c;c)}\,(I\wedge) \qquad \dfrac{\Diamond(a;b),c\vdash\Diamond(a;b) \quad \dfrac{a,c\vdash c \quad b,c\vdash b}{\Diamond(a;b),c\vdash\Diamond(c;b)}\,(\Diamond I1)}{\Diamond(a;b),c\vdash\Diamond(a\wedge c;b)}\,(I\wedge)}{\dfrac{\Diamond(a;b),c\vdash\Diamond((a\wedge c);(b\wedge c))}{\dfrac{\vdots}{(a\|b)\wedge c\vdash(a\wedge c)\|(b\wedge c))}}}\,(I\wedge)$$

ii. $(a \wedge c)\|(b \wedge c) \leq (a\|b) \wedge c$.

$$\dfrac{\dfrac{a,c\vdash c \quad b,c\vdash c}{\Diamond(\natural(a,c);\natural(b,c))\vdash c}\,(\Diamond I2) \qquad \dfrac{\dfrac{a,c\vdash a \quad b,c\vdash b}{\Diamond(\natural(a,c);\natural(b,c))\vdash\Diamond(a;b)}\,(\Diamond I1)}{\Diamond(\natural(a,c);\natural(b,c))\vdash a\|b}\,(I\|)}{\dfrac{\Diamond(\natural(a,c);\natural(b,c))\vdash(a\|b)\wedge c}{\dfrac{\vdots}{(a\wedge c)\|(b\wedge c)\vdash(a\|b)\wedge c}}}\,(I\wedge)$$

($\|$ 2) i. $\neg(a\|b) \leq \neg a\|\neg b$ We slightly abbreviate the proof:

$$\dfrac{\dfrac{\dfrac{\Diamond(a;b)\vdash a\|b}{\Diamond(\flat(\neg a);\flat(\neg b))\vdash\flat(\neg(a\|b)}}{\flat(\neg a\|\neg b)\vdash\flat(\neg(a\|b))}}{\neg(a\|b)\vdash\neg a\|\neg b}$$

ii. $\neg a\|\neg b \leq \neg(a\|b)$ is parallel.

(ass) Straightforward.
(inf) Consider the following (abbreviated) proof for $a \wedge b \leq a\|b$:

$$\dfrac{\dfrac{a,b\vdash a \quad a,b\vdash b}{a,b\vdash\Diamond(a;b)}\,(\Diamond I3)}{a\wedge b\vdash a\|b}$$

$a\|b \leq a \vee b$ can be proved similarly, but already follows algebraically.

(mon) $a\|b \leq (a \vee c)\|(b \vee d)$ is easy to derive from $a \vdash a \vee c$, $b \vdash b \vee d$ and ($\|$I1).

$\square$

So we have a completeness result, following by the standard argument: if a sequent is valid in all algebras, it is in particular valid in **Linda**, the term algebra, hence it is derivable in the calculus.

Theorem 4. **UDA** $\models \Gamma \vdash \Delta$ *if and only if* $\Vdash_{\mathsf{AL}} \Gamma \vdash \Delta$.

5 Conclusion and Further Work

We have presented the logic AL, which is an extension of the classical propositional calculus with an additional ambiguity operator. The main achievements have been the following: we have introduced a multi-sequent calculus which embeds both classical and a substructural logic. For reasons of space, we could not dwell on proof-theoretic properties of this logic, but we conjecture that many structural rules are admissible. What is most interesting is decidability of the calculus, a problem we did not treat in this paper. We rather focussed on the semantics of the calculus: we provided the algebraic semantics by means of universal distribution algebras, which we proved to be sound and complete.

Of course, this work is only preliminary: the exact properties of AL and **UDA** have not even been discussed. Still, we hope that this work shows on the conceptual side that the relation between logic and ambiguity is not an entirely negative one, and that we can effectively reason with ambiguous information. On the formal side, we think that multi-contexts and the extension of classical logic by substructural connectives is a very interesting field in the study of logic, which to our knowledge has yet attracted little attention.

References

1. Atlas, J.D.: Philosophy without Ambiguity. Clarendon Library of Logic and Philosophy. Clarendon Press, Oxford (1989)
2. Dyckhoff, R., Sadrzadeh, M., Truffaut, J.: Algebra, proof theory and applications for a logic of propositions, actions and adjoint modal operators. Electr. Notes Theor. Comput. Sci. **286**, 157–172 (2012)
3. Egg, M.: Semantic underspecification. Lang. Linguist. Compass **4**(3), 166–181 (2010)
4. Kracht, M.: Mathematics of Language. Mouton de Gruyter, Berlin (2003)
5. Lahav, O., Avron, A.: A unified semantic framework for fully structural propositional sequent systems. ACM Trans. Comput. Log. **14**(4), 27:1–27:33 (2013)
6. Lambek, J.: Cut elimination for classical bilinear logic. Fundam. Inform. **22**(1/2), 53–67 (1995)
7. Maddux, R.: Relation Algebras. Elsevier, Amsterdam (2006)
8. Montague, R.: The proper treatment of quantification in ordinary English. In: Hintikka, J., Moravcsik, J.M.E., Suppes, P. (eds.) Approaches to Natural Language, pp. 221–242. Reidel, Dordrecht (1973)
9. Negri, S., von Plato, J.: Structural Proof Theory. Cambridge University Press, Cambridge (2001)
10. Pinkal, M.: Logic and Lexicon: The Semantics of the Indefinite. Kluwer, Dordrecht (1995)

11. Poesio, M.: Semantic ambiguity and perceived ambiguity. In: van Deemter, K., Peters, S. (eds.) Semantic Ambiguity and Underspecification, pp. 159–201. CSLI Publications, Stanford (1994)
12. Restall, G.: An Introduction to Substructural Logics. Routledge, New York (2008)
13. Sennet, A.: Ambiguity. In: Zalta, E.N. (ed.) The Stanford Encyclopedia of Philosophy, Spring 2016 edn. Metaphysics Research Lab, Stanford University (2016)
14. Wurm, C., Lichte, T.: The proper treatment of linguistic ambiguity in ordinary algebra. In: Foret, A., Morrill, G., Muskens, R., Osswald, R., Pogodalla, S. (eds.) FG 2015-2016. LNCS, vol. 9804, pp. 306–322. Springer, Heidelberg (2016). https://doi.org/10.1007/978-3-662-53042-9_18

Advantages of Constituency: Computational Perspectives on Samoan Word Prosody

Kristine M. Yu(✉)
University of Massachusetts Amherst, Amherst, MA 01003, USA
krisyu@linguist.umass.edu

Abstract. In this paper, we computationally implement and compare grammars of Samoan stress patterns that refer to feet and that refer only to syllables in Karttunen's finite state formalization of Optimality Theory, and in grammars that directly state restrictions on surface stress patterns. The grammars are defined and compared in the high-level language of xfst to engage closely with specific linguistic proposals. While succinctness (size of the grammar) is not affected by referring to feet in the direct grammars, in the OT formalism, the grammar with feet is clearly more succinct. Moreover, a striking difference between the direct and OT grammars is that the OT grammars suffer from scaling problems.

Keywords: Phonology · Stress · Finite state · Optimality theory

1 Introduction

A substantial body of work in theoretical phonology suggests that referring to phonological constituents—units such as feet, prosodic words and higher-level constituents—can capture generalizations in phonological patterns [15,35,43,45]. But other work in theoretical phonology offers alternative ways to capture the same kinds of generalizations [28,29,47]. And strikingly, computational descriptions of phonological patterns have revealed strong structural universals without referring to constituents at all [17,18]. Thus, while "we [phonologists] tend to help ourselves to prosodic domains without further comment" [46], there is in fact a puzzle here—*Do constituents make phonological grammars more succinct?*—which hasn't yet been carefully investigated computationally.[1] The work here is an initial effort to begin to fill this gap.

In this paper, we implement and compare phonological grammars of stress patterns in Samoan monomorphs to assess whether grammars referring to feet are more succinct than grammars that refer only to syllables. Succinctness comparisons between grammars of the same ilk have appeared in [5,8,14,33,38,44], among other work. We define all grammars in xfst [4], software for computing with finite

[1] Some computational work has defined phonological patterns in terms of tiers [6,9,30] from autosegmental theory [12], but tiers aren't properly nested like constituents, e.g., it's generally assumed that feet don't straddle prosodic words.

A. Foret et al. (Eds.): FG 2017, LNCS 10686, pp. 105–124, 2018.
https://doi.org/10.1007/978-3-662-56343-4_7

state machines, and our comparison is a 2×2 experimental design because we compare grammars with and without feet in two formalisms: (a) Karttunen's finite state Optimality Theory (OT) [27], which maps underlying forms to surface forms without an intermediate mapping to violations, by using a special composition operator "lenient composition", and (b) a "direct" approach which directly regulates the surface patterns, e.g., [17,22]. Our work thus addresses not only the role of constituents in phonology, but also the advantages and disadvantages of different formalisms. Our method of implementation and comparison is designed to engage closely and concretely with linguistic analyses and empirical data. The goal here is not to make general, abstract claims, but to question if a particular phenomenon motivates prosodic constituents, cf. case studies that examine where phonological and syntactic patterns fall in the Chomsky hierarchy. In doing this, we carefully state very specific assumptions and claims guided by proposals by phonologists about the phenomenon to gain a clearer understanding of the phenomenon and the proposals. The code implementing and testing the grammars is available at https://github.com/krismyu/smo-constituency-feet.

The rest of the paper is structured as follows: the remainder of this introductory section operationalizes *Do constituents make phonological grammars more succinct?* (Sect. 1.1) and describes the simplified language of stress patterns of Samoan that our grammars are designed to capture (Sect. 1.2). We describe the four grammars in Sect. 2: the direct account with feet (Sect. 2.2), the direct account with syllables only (Sect. 2.3), the OT account with feet (Sect. 2.4), and the OT account with syllables (Sect. 2.5). The discussion follows in Sect. 3, and we conclude with Sect. 4.

1.1 Operationalizing the Research Question

In this section, we define the concepts in our research question: phonological grammars, constituents, and succinctness.

Phonological Grammars and Constituents. In Formal Language Theory, a grammar is defined by a finite alphabet of terminal symbols, a finite set of non-terminal categories (which shares no members in common with the alphabet), a finite set of rewrite rules, and a start symbol that initiates the derivation [7,41]. The set of strings that can be derived by the grammar is defined to be the language derived by the grammar. In a tree derived by the grammar, nodes are labeled with categories, and a set of nodes form a constituent if they are exhaustively dominated by a common node.

The nature of restrictions on the structure of rewrite rules determines the structural complexity of the grammar. A standard definition of a (right) regular grammar says that it is a grammar where the rules are restricted to the form $A \rightarrow aB$, and $A \rightarrow \epsilon$, where A and B are non-terminal categories and a is a terminal symbol. This restriction results in grammars where the only constituents of length >1 are suffixes. But prosodic constituents in phonological theory include both prefixes and suffixes. For instance, given an alphabet of light and heavy syllables $\Sigma = \{L, H\}$, suppose we defined a (right) regular grammar that could derive the string $LLHLL$. Then the suffix-constituents derived by the grammar would never be able to pick out the initial LL or the medial H as feet.

A regular language is one that can be derived by a regular grammar. It has been shown that (almost) all phonological patterns are regular [24,26]. We are left with an apparent contradiction: if phonology has constituents that are both prefixes and suffixes, then how is it that phonology is regular? The critical point is that it's the *language*—the set of admissible surface patterns—that's been shown to be regular in phonology, not the grammar that could derive it. There are infinitely many grammars that can define the same language; supra-regular grammars—with fewer restrictions on rewrite rules than regular ones—can define regular languages. A language that can be defined by a grammar with suffix-constituents can also be defined by one with non-suffix constituents. This paper shows how we might assess which kind of grammar is, in a certain sense, better.

We define the four accounts as regular transductions. We start with a transduction that overgenerates: it marks up input sequences of light and heavy syllables with all possible stress patterns (Gen, defined in Sect. 2.1). From this point on, the syllable-based grammars are defined with identity transductions (i.e., acceptors). However, the foot-based "grammars"[2] require an additional (non-identity) transduction because they also mark up the stressed sequences with boundary symbols indicating left and right foot edges. Here we are coding constituents into the state, not in the derivation tree. An approach that codes constituents into the state can provide an exact account if the bound on tree depth required is finite, and here we are only coding feet—constituents up to depth 1.

We implement (non-suffix) constituents into the state (with regular transducers), rather than into the derivation (with supra-regular acceptors) to keep the definition of the grammars close to those that phonologists use. Linguists have characterized a wealth of phonological patterns with SPE-style rewrite rules, introduced in [8], and also with optimality-theoretic (OT) constraints [36]. [26] showed that the expressivity of SPE-style phonological characterizations is equivalent to that of regular transductions (provided that cyclic application of rules is not permitted). And [10] showed that an OT characterization has the expressivity of a regular transduction if the mapping from input to possible output forms (Gen) and the constraints (Con) are regular, and the number of violations a constraint can assign is finitely bounded; moreover, OT limited in such a way is sufficient for capturing analyses proposed by phonologists, except for analyses with gradient constraints (which can assign unboundedly many violations). Thus, the analytical tools that phonologists use—keeping the hedges mentioned above in mind—have the expressivity of regular transductions.

While phonologists may work with the power of regular transductions, they do not define phonological grammars by specifying transductions in the standard

[2] Because they are defined with non-identity transductions, the foot-based "grammars" are not grammars as defined by Formal Language Theory. But the phonological literature calls phonological transductions—input-output mappings from underlying forms to surface forms—like these "grammars", and we'll follow that convention.

way, by listing states and transitions. Instead, they specify them at a high-level. We can do this too, by writing the grammars in xfst [4]. This language was carefully designed by linguists to make it easy for us to express generalizations in the high-level language which are hard to express or detect at the level of a regular grammar or a finite state machine. It includes pre-defined complex operators to give us a high-level notation for regular transducers. For example, we can write SPE-style rules using replacement rules with the syntax A -> B || L _ R, where we simply need to specify the focus, change, and the structural description: this rule clearly does not meet the form $A \rightarrow aB$. xfst allows us to define our own operators and units, too, e.g., feet, and it compiles our high-level grammars to machine-level finite state transducers. Since xfst grammars are compiled into standard finite state representations, an xfst definition establishes that the phenomenon described is regular (see [11]) and gives us a common formalism in which we can define all four grammars. It is not possible to define "standard OT" [37] fully in xfst: we could define Gen, which generates the set of candidate outputs, and the assignment of violation marks according to the set of violable constraints in Con—but not the non-regular Eval relation for computing the optimal candidate, since the number of states required to define Eval can't be bounded [10]. However, we can define Karttunen's finite state formulation of OT [27] in xfst, as it avoids Eval by mapping underlying forms directly to surface forms with "lenient composition".

Succinctness. Having xfst as a common formalism for defining all four grammars allows us to make a controlled comparison of the succinctness of the grammars. We define the succinctness of a grammar as its size—the number of symbols it takes to write it down (in xfst), under the conventions specified in Sect. 2. We define size over the high-level xfst grammar rather than at the machine level because it's the high-level language that we can express and detect generalizations in; the machines that xfst compiles are big and redundant by comparison. Defining size in this way over a high-level language follows other linguistic work, e.g., [5,8,33,44]. And xfst is a reasonable choice for the high-level language because it was designed by linguists to make it easy to state linguistic generalizations, and not, say, tailor-made to prefer feet over non-feet.

Our metric for succinctness can be thought of a special case of minimum description length (MDL) [39], relativized to the descriptive notation provided by xfst. MDL as a metric for succinctness balances the minimization of the size of the grammar, which favors simple grammars that often overgenerate, with minimization of the size of the data encoded by the grammar, which favors restrictive but often overly memorized grammars. The alphabet over which the grammars are defined (primary, secondary, and unstressed light and heavy syllables) is constant across grammars.[3] Since all grammar definitions are expressed in the same language, we don't need to translate them into some common language like bit

[3] The foot-based accounts also introduce (,), and X as symbols, where X is an unparsed syllable, but: (i) it's not clear these should be included in the alphabet since they come in only in the calculation of stress, (ii) if they are included, they make a negligible difference.

representations; we can simply measure xfst grammar size with symbol counting. The data is also the same across the comparisons (the set of stress patterns in words elicited from linguistic consultants, plus some predicted ones up to 5 syllables; we have empirical data for monomorphs only up to 5 syllables). Moreover, as we show in Sect. 3, all the grammars admit exactly the same set of stress patterns up to 5 syllables (with one exception). Thus, the MDL metric, relativized to the descriptive resources of xfst, reduces to the size of the grammar. That is, the size of their encodings of the sequences up to 5 syllables is exactly the same, since the possibilities allowed by the grammars in that range is identical. We accordingly consider just the size of the four grammars, all expressed in the common xfst formalism.

1.2 Description of the Language Little Samoan

Samoan stress presents a good case study for a first comparison of the succinctness of grammars with and without feet. We can engage closely with the literature and empirical data because a recent detailed foot-based OT analysis of Samoan stress based on a rich set of elicited words is available [47]. [47]'s analysis also extends to morphologically complex words parsed into multiple prosodic words. In future work, we plan to extend the work here by comparing grammars with higher-level prosodic constituents than feet and modeling parsing at the syntax-phonology interface.

We define Little Samoan (LSmo), a language of strings of syllables marked for stress and weight, as a simplified version of the description of Samoan stress in monomorphs provided in [47]. LSmo is defined over light and heavy syllables rather than segments, and thus ignores complications from diphthongization and the interaction of stress with epenthesis. In [47]'s description, Samoan outputs LL for HL-final words to avoid heavy-light (HL) feet, and also presumably for L (content) words, to satisfy minimal word constraints. Since our OT grammars don't change the syllable weights in the input, we model this in OT by mapping LL and HL-final inputs to a special Null symbol denoting a null output [37]. Our direct models simply define transductions that don't accept HL and LL.

The basic primary stress pattern in LSmo (and Samoan) is moraic trochees at the right edge [47, (4)], e.g., la('vaː) 'energized', ('manu) 'bird', i('ŋoa) 'name'; exactly like the well-known stress system of Fijian [15, Sect. 6.1.5.1], "if the final syllable is light, main stress falls on the penult; if the final syllable is heavy, main stress falls on the final syllable". Secondary stress in LSmo is almost like in Fijian, where "secondary stress falls on the remaining [non-final] heavy syllables, and on every other light syllable before another stress, counting from right to left" [15, Sect. 6.1.5.1, p. 142], e.g., (ˌmaː)(ˌloː)('loː) 'rest' [47, (7)]. However, LSmo has an initial dactyl effect: initial LLL sequences are initially stressed, e.g., ('mini)si('taː) 'minister' (cf. Fijian mi(ˌnisi)('taː)), ('temo)ka('lasi) (cf. Fijian pe(ˌresi)('tend͡i) 'president').[4]

[4] This ignores [47]'s evidence from LLLLL loan words showing that an initial weak-strong-weak (WSW) pattern can occur if the first vowel in the word is epenthetic.

[47] provides data on only two monomorphs that are longer than 5 moras: [(ˌmaː)(ˌloː)(ˈloː)] and a 7-mora all light loanword for Afghanistan. As noted in [47, p. 281, fn. 2] the consultant produced (ˌʔafa)(ˌkani)si(ˈtana), while a true initial dactyl pattern would yield [(ˌʔafa)ka(ˌnisi)(ˈtana)]. Little data is available for longer words in Fijian, too, and there is variability [te(ˌreni)(ˈsisi)(taː)] 'transistor' vs. [(ˌkeː)(ˌmisi)ti(ˈriː)] 'chemistry', which may be due to faithfulness to stress in the source word and a dispreference for stressing epenthetic vowels, as in Samoan [15, p. 144, (44b, 47)]. Due to the lack of data on longer words, we limit testing empirical coverage of LSmo to monomorphs of 5 syllables (although of course our grammars can accept input strings of arbitrary length). Also, we allow a more general distribution of dactyls, which permits both initial or medial dactyls. This allows both (ˌLL)(ˌLL)L(ˈLL) and (ˌLL)L(ˌLL)(ˈLL), and predicts, for example, that HLLLH may be (ˌH)(ˌLL)L(ˈH) or (ˌH)L(ˌLL)(ˈH).

2 Four Grammars for Little Samoan

In this section, we describe our grammars for LSmo: the foot-based direct account (Sect. 2.2), the syllable-based direct account (Sect. 2.3), the foot-based OT account (Sect. 2.4), and the syllable-based OT account (Sect. 2.5). We define the transductions in the grammar in xfst, with symbol counts in square brackets to the right of the command. All of the xfst expressions we use are definitions, which get compiled into transducers associated with a variable. These have the syntax define *variable-name xfst-expression.* Our conventions for writing xfst expressions and counting symbols are as follows: (a) xfst expressions are delimited by square brackets in a define command. (b) Auxiliary terms are defined for any expression that appears more than once in the grammar. (c) A conjunct or disjunct longer than one symbol is enclosed in brackets. (d) A semicolon ending a line counts as a symbol; spaces do not count. (e) A number or a character enclosed in double quotes, e.g., "(", counts as one symbol. (f) Each variable name, command, operator, and atomic expression counts as a single symbol, i.e., define, Heavy, WeakLight, etc., ?, *, +, ^, .#., [,], (,), |, &, ~, \, $, ->, ,, =>, ->@ _, ..., .o., .O. (g) Symbols used to define Gen (constant across the grammars) don't contribute to the symbol count.

Each of the four grammars is implemented as a cascade of transducers and all four share the same basic schematic architecture:

```
addMarkup .o. defineSyllableAndFootTypes .o. enforceSurfaceRestrictions
```

All accounts begin with the transduction Gen to mark up input sequences of light and heavy syllables with stress, as described in Sect. 2.1. The foot-based accounts additionally add markup to indicate foot edges. These markup transductions are the only non-identity transductions in the implementations. They overgenerate stress patterns.[5] These are then filtered in subsequent transductions that define

[5] We set up the initial generation of stress patterns like Gen in Standard OT [36, Sects. 2.2, 5.2.3.3] for the direct accounts as well as the OT accounts. We do this for convenience; we could also generate in some other way for the direct accounts.

sub-types of syllables (and feet in foot-based accounts) and then restrict stress patterns in terms of these syllable (and foot) types. The difference between direct and OT accounts shows up in the way the surface restrictions are expressed in the grammar: the OT accounts are limited to define the restrictions as OT constraints used by phonologists, while the direct accounts are not.

2.1 Preliminaries: Adding Stress Markup with Gen

Let us call the transduction that generates all possible sequences of syllables marked with degree of stress and weight Gen. We define Gen (4) as the composition of Input (1), SWParse (2), and ElevateProm (3). Input generates Σ^* over the alphabet of light (L) and heavy (H) syllables, $\Sigma = \{L, H\}$. Then, SWParse marks the degree of stress on a syllable by inserting labeled brackets around each syllable,[6] (see Parse in [27] and also [4, p. 68]), e.g., the input L has the output S[L] (strong/stressed), W[L] (weak/unstressed). Finally, ElevateProm optionally replaces any strong syllable S[ ] with a primary stressed syllable, P[ ], so that S[ ] now stands for secondary stress. As an example, input LL is mapped to { P[L]P[L], P[L]W[L], P[L]S[L], W[L]P[L], W[L]W[L], W[L]S[L], S[L]P[L], S[L]W[L], S[L]S[L]}.

```
define Input ["L" | "H"]*; [9]                                    (1)
define SWParse [ ? -> [ "S" "[" | "W" "[" ] ... "]"]; [15]        (2)
define ElevateProm ["S"(->)"P"]; [10]                             (3)
define Gen [ Input .o. SWParse .o. ElevateProm ]; [10]            (4)
```

2.2 Direct Account with Feet

ParseFoot (6) parses the output from Gen into feet, marking it up with boundary symbols; it refers to auxiliary terms for heavy [H] and light [L] syllables, Heavy and Light (5). We restrict a foot to being bimoraic: either a LL or a H, so ParseFoot wraps parentheses pairs around any LL or H, e.g., (P[L]W[L]), regardless of the stress pattern. We then define types of feet to express restrictions on stress patterns. Foot defines a foot as string of non-parentheses enclosed in parentheses (7); PrimaryFoot defines a primary stressed foot as a string accepted by Foot that also includes P (8), and WeakLight defines a weak light syllable (9) using Light. We define trochaic feet with Trochee (12), which accepts a strong-weak LL sequence LLFoot (10), or a strong H HFoot (11). The sequence \"W" in (10) and (11) indicates the negation of the character W, i.e., any character but W such as P or S, which mark strong syllables.

With parsing the input into feet and definitions of types of feet behind us, the payoff comes as we can express LSmo's restrictions on stress patterns in terms of feet. TrocheesOnly (13) forces feet to be trochaic: it only accepts strings

[6] We assume that syllable splitting feet do not occur [15, Sect. 5.6.2, p. 121].

that are sequences of trochees that may be interspersed with unparsed syllables (weak lights), e.g., it winnows down the parses for LL to just (S[L]W[L]) and (P[L]W[L]). Additionally, PrimaryFootRight accepts only strings which terminate in a foot bearing primary stress and whose final foot is not preceded by any other primary stresses (14), e.g., eliminating (S[L]W[L]). Note that this transduction eliminates lone Ls and HL-final strings in the language, since they do not have parses with final primary stressed feet. Finally, we implement the "initial dactyl effect" in LSmo with InitialDactyl (15), which forbids a sequence of a weak light followed by a LL foot at the beginning of the word, if the LLL sequence is non-final, i.e., followed by at least one character (?+). This yields the SWW-initial output (S[L]W[L])W[L](P[H]) for LLLH sequences ([(ˌmini)si(ˈtaː)], [47, (8)]), but a WSW pattern W[L](P[L]W[L)] for LLL sequences [i(ˈŋoa)], [47, (4)]. This also allows outputs for HLLLH and 7Ls with medial dactyls. The final transduction going from all possible sequences of stressed and weighted syllables to only those in LSmo composes the foot parser with the restrictions on words in terms of feet (16).[7] All together, excluding Gen, the grammar (5–16) costs 141 symbols.

```
define Heavy [ "[" "H" "]"]; define Light [ "[" "L" "]"]; [16] (5)
define ParseFoot [ [ [ ? Light ]^2 | [ ? Heavy ] ] ->"(" ... ")"]; [22] (6)
define Foot ["("\. ["(" | ")"]]* ")"]; [16] (7)
define PrimaryFoot [ Foot & $["P"] ]; [11] (8)
define WeakLight ["W" Light ]; [7] (9)
define LLFoot ["(" \"W" Light WeakLight ")"]; [11] (10)
define HFoot ["(" \"W" Heavy ")"]; [10] (11)
define Trochee [ LLFoot | HFoot ]; [8] (12)
define TrocheesOnly [ Trochee | WeakLight ]*; [9] (13)
define PrimaryFootRight [ \"P"* PrimaryFoot ]; [9] (14)
define InitialDactyl ~[ WeakLight LLFoot ?+ ]; [10] (15)
define LSmoDirFt [ ParseFeet .o. TrocheesOnly
    .o. PrimaryFootRight .o. InitialDactyl ]; [12] (16)
```

2.3 Direct Account Referring to Syllables only

The definition of Gen in this account is the same as in Sect. 2.2, but then we state restrictions on stress patterns in terms of syllables, rather than over feet. In addition to the previously defined auxiliary terms Heavy and Light (5), and WeakLight (9), we also define: PrimaryLight and SecondaryLight, a primary P[L] and secondary stressed light syllable S[L] (17); StressedSyll, a syllable of any weight that is not weak (18); and W2, a sequence of two weak lights W[L]W[L] (a lapse) (19).

[7] LSmoDirFt can also be composed with a transduction that replaces W in unparsed syllables with X, to match notation for the OT footed account in Sect. 2.4.

Restrictions on the distribution of secondary and weak lights are more complex and are expressed as a series of cases. StressSLight (22) restricts a secondary light to be followed by a non-final weak light (so a penult light cannot receive secondary stress). In addition, a secondary light must be string-initial, preceded by a lapse W[L]W[L], or preceded by a S[L]W[L], i.e., in terms of feet, start a new foot. We must further restrict the position of weak lights, because the transducer that is the intersection of (20), (21), and (22) admits strings with lapses anywhere, e.g., it accepts both P[L]W[L] and W[L]W[L] from the set of sequences generated from input LL. RestrictLapse (23) restricts a lapse W[L]W[L] to be preceded by a secondary light and followed by a stressed syllable, allowing a lapse just in case it is in a dactyl S[L]W[L]W[L] and not string-final. The intersection of StressPLight, StressSLight eliminates a lone stressed L, since StressPLight and StressSLight restrict stressed lights to being non-final. But RestrictLapse does not eliminate a lone weak L. Moreover, no transduction thus far eliminates HL-final sequences.[8] Thus, we must define transducers to ban HL-final sequences and lone Ls: NoFinHL and NoLoneL (24). All together the grammar costs 145 symbols.

```
define PrimaryLight ["P" Light]; define SecondaryLight ["S" Light]; [14]   (17)
define StressedSyll [ \"W" [ Heavy | Light ] ]; [12]   (18)
define W2 [ WeakLight WeakLight ]; [7]   (19)
define StressHeavy [ Heavy =>"P" _ .#., "S" _ ]; [13]   (20)
define StressPLight [ PrimaryLight => _ WeakLight .#. ]; [10]   (21)
define StressSLight [SecondaryLight =>
[.#. | W2 | [StressedSyll WeakLight] | Heavy] _ WeakLight ? ]; [19]   (22)
define RestrictLapse [ W2 => SecondaryLight _ StressedSyll ]; [10]   (23)
define NoFinHL ~[?* ? Heavy ? Light]; [10] define NoLoneL ~[Light] ; [7]   (24)
define LSmoDirSyll [ Gen .o. [ StressHeavy & StressPLight &
StressSLight & RestrictLapse ] .o. NoFinHL .o. NoLoneL ]; [20]   (25)
```

2.4 Karttunen OT with Feet

Our constraint set for this account is a subset of the constraints used in [47]; we have removed constraints that are only relevant for segments, morphologically complex words and multiple prosodic words. The partial ranking was computed with OTSoft[9] [16] based on monomorphemic candidates used in [47] and is: *Stratum 1* (i) FootBinarity (FtBin) A foot must contain exactly two moras. (ii) RhythmType=Trochee (RhType=Trochee) A foot must have stress on its initial

[8] HL-final sequences are allowed in [17]'s acceptor for Fijian stress (http://phonology.cogsci.udel.edu/dbs/stress/language.php?id=109), based on [15]'s basic description of Fijian stress, but [15, p. 145, Sect. 6.1.5.2]'s more detailed description suggests that they should not be accepted.

[9] All OTSoft input and output files are in the github repository.

mora, and its initial mora only. (iii) Align(PWd,R; 'Ft,R) (Edgemost-R) The end of the prosodic word must coincide with the end of a primary-stressed foot; *Stratum 2* (i) Parse-σ Every syllable must be included in a foot. (ii) Align(Pwd;L,Ft,L) The beginning of the prosodic word must coincide with the beginning of a foot. Constraints in a earlier stratum are ranked higher than those in a later one, but constraints within a stratum are not ranked with respect to one another.

```
define MarkUnparsed [ "W" (->) "X" ]; [10]  (26)
define FtParse [ [ \"X" [ Heavy | Light ] ]+ -> "(" ... ")" ]; [19]  (27)
define GenFt [ Gen .o. MarkUnparsed .o. FtParse]; [10]  (28)
define Culminativity [ $.P ]; [8]  (29)
define NullFinHL [  [?* Heavy [?]^<4 Light (")") ] ->@ "Null" ]; [27]  (30)
define NullLoneL [ [ ("(") ? Light (")") ] -> "Null" ]; [21]  (31)
define Unparsed [ "X" "[" ? "]" ]; [8]  (32)
define FtBinH [ "(" ? Heavy ")" ]; [7] define FtBinLL [ "("[ ? Light]^2 ")" ]; [13]  (33)
define FtBin [ FtBinH | FtBinLL | Unparsed ]*; [11]  (34)
define Stressed [ "S" | "P" ]; [8]  (35)
define RhTypeTrocheeH [ [ Heavy ")" ] => "(" Stressed _ ]; [13]  (36)
define RhTypeTroLL [ "[" "L" => "(" Stressed _ ,  "]" "W" _ , "X" _ ] ; [18]  (37)
define RhTypeTrochee [ RhTypeTrocheeH & RhTypeTrocheeLL ]; [8]  (38)
define ParseSyll ~[$"X"]; [8]  (39)
define ParseSyll1 ~[[$"X"]^>1]; [13] define ParseSyll2 ~[[$"X"]^>2]; [13]  (40)
define AlignWdLFtL [ "(" ?* ]; [8]  (41)
define LSmoMonoFtOT [ GenFt .O. NullFinHL .O. NullLoneL .O. Culminativity
.O. FtBin .O. RhTypeTrochee .O. EdgemostR .O. ParseS .O. ParseS1 .O. ParseS2
.O. ParseS3 .O. ParseS4 .O. ParseS5 .O. AlignWdLFtL ]; [28]  (42)
```

We compute constraints referring to a prosodic word edge with respect to the edge of the input string since the input never contains more than a single prosodic word. With the exception of Edgemost-R, our constraint definitions are identical to those in [47, (5), (12)]. Some constraints, called categorical, assign multiple violations to a candidate iff there are multiple places where the constraint is violated in the candidate, e.g., Parse-σ (39). Other constraints, called gradient, "measure the extent of a candidate's deviance from some norm", and can assign multiple violations even if there is a single locus of violation in the input [31, p. 75]. [47] computes Align(Pwd;L,Ft,L) as a categorical constraint, e.g., 1 violation for *[sika('lamu)] 'scrum'. However, Edgemost-R in [47] is computed gradiently, e.g., 2 violations for *[('piː)niki]. We define it instead as categorical: since it's undominated in our constraint set, whether it is assessed categorically or gradiently makes no difference. We can paraphrase the constraint definition as "assign a violation for every PWd where there exists a primary-stressed foot such that the right edge of the PWd and the right edge of the primary-stressed foot do not coincide" [32]. As we'll discuss in Sect. 3, whether a constraint is categorical vs. gradient markedly impacts the succinctness of the grammar in Karttunen's formalism, as does whether a constraint can be multiply or only singly violated.

We define the candidate set in the OT footed account with GenFt (28) as the composition of Gen (4), MarkUnparsed (26), and FtParse (27). MarkUnparsed optionally replaces W[ ] with X[ ] to mark syllables unparsed into feet. FtParse parses the input into feet by wrapping parentheses around a non-empty sequence of syllables that are not unparsed, e.g., X[L](P[L]W[L]). We define three undominated constraints not included in the OTSoft ranking: Culminativity (29), NullFinHL (30), and NullLoneL (31). We define Culminativity (every word must contain exactly one primary stress) as a constraint rather than a property of Gen to keep Gen constant across grammars. NullFinHL and NullLoneL map HL-final and L inputs to Null. These definitions allow the HL-final and L inputs to be footed or unfooted and use directed replacement operators [4, p. 73]. For the -@> operator, replacement strings are selected right to left, and only the longest match is replaced.

We then define constraints on feet. FtBin (34) restricts footed sequences to be any sequence of heavy feet and LL feet (FtBinH, FtBinLL, (33)), and unparsed syllables (Unparsed, (32)). We define RhyTypeTrochee (38) as the conjunction of RhyTypeTrocheeH (36) and RhyTypeTrocheeLL (37). RhyTypeTrocheeH restricts a heavy syllable followed by a parentheses, i.e., a footed H, to be preceded by a parentheses and S (secondary) or P (primary): a footed H must be stressed. RhyTypeTrocheeLL restricts a light syllable to be foot-initial and stressed (defined with Stressed (35)), or non-initial in a foot and weak, or unparsed. It accepts SWW dactyls. EdgemostR (not shown) is identical to PrimaryFootRight (14).

ParseSyll (39) must be implemented as a family of constraints because it can be multiply violated; we discuss this further in Sect. 3. Each ParseSyllN constraint in the family restricts the input string to have no more than N substrings containing X, e.g., $N = 1$ in ParseSyll1 (40). We follow the implementation in [27, pp. 10–11][10] We must impose some finite k-bound on the family; here we set $k = 5$ since the range of patterns we want to account for are only as long as five syllables. AlignWdLFtL (41) states that the beginning of the input string must coincide with the beginning of a foot and then may be followed by any string. The final transduction LSmoMonoFtOT is defined as a "lenient composition" (42).[11] [27] defines this (.O.) to be a special form of composition where input strings are held back from being eliminated to keep the set of output candidates from becoming empty. The order of "lenient" composition is important: the higher ranked a constraint is, the earlier it must enter the composition. Within a stratum, order of composition doesn't matter. In total the OT footed grammar costs 306 symbols, including 73 from the ParseSyll family (as well as 16 from previously defined Light and Heavy (5)).

---

[10] But there's an inconsistency in [27]'s definitions of Parse; it should be defined as ~$["X["]; and not ~[$"X["]; in Figs. 8 and 16.

[11] We abbreviate ParseSyll as ParseS for space; see github repository for definitions of ParseSyllN for $N > 2$.

2.5 Karttunen OT with Syllables only

The grammar using Karttunen OT with syllables only is by far the biggest grammar: it includes not only categorical constraints that can be multiply violated, but also gradient alignment constraints. There are few OT analyses of stress patterns that are not based on feet, i.e., "grid-based" [3,13,25], and we drew on constraints from them, but failed to generate only the allowed stress patterns up to 5 syllable words without introducing ad-hoc constraints that referenced feet without naming them. The partial ranking was computed with OTSoft [16] and is: *Stratum 1.* (i) WeightToStress (WSP) A heavy syllable must be stressed. (ii) NonfinalityL A word-final light syllable must be unstressed. (iii) NoLapseFollowingHeavy A heavy syllable musn't be followed by two unstressed syllables. (iv) NoInitialWS A word musn't begin with an unstressed-stressed sequence. *Stratum 2.* Align(x2,R,x0,PWd) Assign a violation for every grid mark of level 2 that doesn't coincide with the right edge of level 0 grid marks in a prosodic word. *Stratum 3.* *Clash Assign a violation for every sequence of two stressed syllables. *Stratum 4.* *Lapse Assign a violation for every sequence of two unstressed syllables. *Stratum 5.* Align(x1;L,x0,PWd) Assign a violation for every grid mark of level 1 that doesn't coincide with the left edge of level 0 grid marks in a prosodic word.

The xfst grammar for syllable-based Karttunen OT is shown in (43)–(57). After first introducing some auxiliary terms, we define the undominated constraints: WeightToStress (46), NonfinalityL (nonfinality restricted to light syllables) (47) [23], and two ad-hoc constraints—NoLapseFollowingHeavy and NoInitialWS. NoLapseFollowingHeavy (48) essentially enforces that a LL sequence after a heavy should be footed, and thus receive stress and allows the general *Lapse constraint to be ranked lower. NoInitialWS (49) essentially bans iambic feet word-initially. To achieve the initial dactyl effect, we use grid-based gradient Align-x constraints drawn from the schema in [13, (2)]. Align(x2,R,x0,PWd) enforces primary stress towards the right, while Align(x1;L,x0,PWd) is necessary to promote SWW initial candidates. While WSP (46) and NoLapseFollowingHeavy may have multiple loci of violation, because they are undominated, we were able to implement them as if they could only be singly violated. However, we had to implement a family of constraints to effectively count multiple violations for *Clash, *Lapse, and Align(x2,R,x0,PWd). We use Culminativity (29) to filter out strings with multiple primary stresses, so we could implement Align(x2,R,x0,PWd) similarly to the ParseSyll family. But the implementation of Align(x1;L,x0,PWd) requires doing arithmetic because the same number of violations could be incurred by multiple stress patterns. We present a selection of the grammar below (see the github repository for the full grammar); Gen (4) is the same as before. Previously defined transductions repeated here (not shown) are: Culminativity (29), Heavy and Light (5), and W2 (19). We also defined Weak (43) and Stressed (44) syllables, and clash S2 (45), two adjacent stressed syllables. Transductions similar to NullFinHL (30) and NullLoneL (31) map HL-final and L inputs to Null (not shown).

The NoNClash family of constraints restricts the input from having N clashes: if an input is not accepted by NokClash, then it is also not accepted for any

$N > k$. The constraints define languages that are in a strict subset relation, like [27]'s ParseSyll family. We implemented NoNClash constraints (for $N = 1, 2, 3, 4$) as conjunctions, because there are multiple stress patterns that can result in the same number of clashes. For instance, an input may have two clashes because it has two nonadjacent clashes, or because it has a sequence of three stressed syllables, see No2Clash (54). The higher N is, the more conjuncts in the definition; specifically, it is 2^{N-1}. Already for No3Clash, we require 4 conjuncts: strings containing a substring of 4 stressed syllables (SSSS, where S stands for stressed syllable), two substrings containing SSS sequences, a substring containing SSS followed by a substring containing SS, and a substring containing SS followed by a substring containing SSS. The definition of NoNLapse is identical to that of NoNClash, but replaces S2 with W2 and Stressed with Weak.[12]

```
define Weak "W" [ Heavy | Light ]; [9]   (43)
define Stressed ["S"|"P"] [ Heavy | Light ] ; [13]   (44)
define S2 [ Stressed Stressed ]; [7]   (45)
define WSP [ Heavy  => \W _ ]; [10]   (46)
define NonfinalityL ~[?* \"W" Light]; [11]   (47)
define NoLapseFollowingHeavy ~[$[Heavy W2]]; [11]   (48)
define NoInitWS ~["W" Light "S" Light ?*]; [12]   (49)
define AnySyll [ ? "[" ? "]" ]; [9]   (50)
define Alignx2R0 [ Primary => _ .#. ]; [9]   (51)
define Alignx2R1 [ Primary => _ [AnySyll]^<2 .#. ]; [15]   (52)
define No1Clash ~[$[S2]]; [10]   (53)
define No2Clash ~[$[Stressed]^3] & ~[[$[S2]]^2]; [25]   (54)
define SWStar [Stressed [Weak]*]; [10]   (55)
define Alignx1L1 ~[AnySyll SWStar]; define Alignx1L2 ~[AnySyll Weak SWStar];   (56)
define Alignx1L3 ~[AnySyll W2 SWStar] & ~[AnySyll Stressed SWStar]; [16]   (57)
```

Similarly, we implement Align(x2,R,x0,PWd) as a categorical constraint, as the family Alignx2RN. We can do this because there is only ever one grid mark of level 2 (primary stress) in our candidates, since we restrict them to be single prosodic words. Each transducer Alignx2RN restricts primary stress to be at most N syllables from the right edge, where the syllables can be of any type (AnySyll (50)). All Alignx2RN transducers for $N > 1$ have the same form as Alignx2R1 (52); the languages accepted by the transducers are in a strict subset relation, and for words only up to 5 syllables, Alignx2RN for $N > 4$ may be omitted.

Align(x1;L,x0,PWd), though, must be implemented as a gradient constraint. There can be multiple grid marks of level 1 (stressed, i.e., not weak) in our

[12] See the github code repository for definitions of No3Clash (61 symbols) and No4Clash (131 symbols) and the NoNLapse constraint family.

candidates. We do it in two parts. First we define the Alignx1LN family, a set of transducers where Alignx1LN restricts the input to have the sum of the distances that stressed syllables are away from the left edge to total N. For example, Alignx1L3 (57) does not accept inputs that are in ?WWSW* (3 violations) or ?SSW* (1 + 2 violations), and Alignx1L5 does not accept inputs that are in ?WWWWSW* (5 violations), ?SWWSW* (1 + 4 violations), or ?WSSW* (2 + 3 violations), where W stands for a weak syllable, S for a stressed syllable, and ? for any character. The family definition refers to auxiliary term SWStar, the language of a stressed syllable followed by any number of unstressed syllables (55). Definitions for $N = 1, 2, 3$ are given in (56, 57).

We then take intersections of the Alignx1LN languages to define languages Alignx1LgM that accept any number of violations less than M. For example Alignx1Lg5 is the intersection of Alignx1L5, Alignx1L6, Alignx1L7, ..., Alignx1Lk, "don't have 5, 6, 7, ... k violations" where k is some finite upper bound. The Alignx1Lg5 language contains S[H]S[L]W[L]P[H], which has a total of $1 + 3 = 4$ violations, but not S[H]W[L]S[L]P[H], which has a total of $2 + 3 = 5$ violations. How high does k need to be? The output S[H]W[L]S[L]S[H]P[H] for HLLHH has $2+3+4 = 9$ violations in total, while the output S[H]S[L]W[L]S[H]P[H] has $1+3+4 = 8$ violations in total. With our constraints, these two candidates have no other difference in their violation profiles. Thus, our transduction should admit the candidate with 8 violations, and not the one with 9. The candidate with 8 violations should be in any Alignx1LgM language where $M > 8$, in particular, in the Alignx1Lg9 language, while the candidate with 9 should not. However, if $k = 8$, and we define Alignx1LgM transducers up to $M = 7$, then Alignx1Lg7 is the intersection of Alignx1L7 and Alignx1L8 "don't have 7 or 8 violations". Then the candidate with 9 violations is in the Alignx1Lg7 language, while the candidate with 8 isn't: the transduction outputs the wrong candidate. To output the correct one, we must have $k \geq 10$, with $M \geq 9$: even for only up to 5-syllable words, we must have $k \geq 10$. Minimally we must include Alignx1L10 in the conjunction that defines Alignx1Lg9. In general, a n-syllable word can have a maximum of $\sum_{i=2}^{n-1} i$ violations of Align(x1;L,x0,PWd) and k must be greater than that sum.

The language derived by the OT syllable grammar has a small difference from the others: while all the other grammars admit two outputs for HLLLH: S[H]S[L]W[L]W[L]P[H] or S[H]W[L]S[L]W[L]P[H], the OT syllable account admits only S[H]W[L]S[L]W[L]P[H]. But [47] doesn't actually include elicited data for HLLLH, so we don't know which pattern(s) our consultants would accept. The striking difference about the OT syllable grammar is in its size. It is much larger than any of the other grammars, and the growth of the size of the grammar increases rapidly with the length of the input string: the definition of just the gradient constraint Align(x1;L,x0,PWd) takes more symbols than any of the other entire grammars, and the definitions of the clash and lapse constraints alone already grow exponentially with the size of the input. Moreover, all of the constraints which can be multiply violated cannot be implemented as finite state transducers, and our implementations approximating these constraints require doing arithmetic and defining constraint families that have the effect of counting up violations to some finite bound. Finally, to just get near-coverage of the data without feet, we needed to define ad-hoc constraints that referenced feet without revealing generalizations in the structural restrictions on stress patterns. And additional exploratory calculations from OTSoft showed that we still need gradient alignment constraints to fit the data, despite including additional categorical constraints from [25]'s Rhythmic Licensing Theory—designed to avoid gradient constraints (see github repository, `otsoft-files/syll/test-with-rlt`).

3 Discussion

We examined the set of possible stress patterns from our grammars by composing our final transducers with an identity transducer that was defined as a disjunction of elicited/expected stress patterns for LSmo up to 5 syllables. We also defined an identity transducer for all possible light-heavy inputs up to 5 syllables and composed that with our final transducers. Then we checked that the set of strings defined by these two compositions was identical for each grammar and across grammars. Our four grammars for LSmo admit exactly the same set of stress patterns up to 5 syllables, with the one exception mentioned above: the OT syllable account admits only one stress pattern for HLLLH.[13] Thus, the MDL metric, relativized to the descriptive resources of xfst, reduces to the size of the grammar, although that's not quite the case for the OT syllable account. The direct accounts were almost the same in size: 145 symbols for the syllable account and 141 symbols for the footed account; the OT footed grammar cost 306 symbols. The small differences between these is insignificant, compared to the qualitatively different character of exponential growth we saw in definition of the OT syllable grammar, with a count in the 1000s. Even if including a battery of constraints from Rhythmic Licensing Theory [25], we found that an OT syllable grammar would still need to include multiply violated clash and lapse constraints and gradient Align constraints; we'd also expect this to be true in general beyond the Samoan case study here, such that the size of OT syllable-based grammars would in general blow up.

Our results show that with a direct account, a grammar referencing feet in the description of stress patterns in LSmo is about as succinct as a grammar that does not. By this metric, one isn't preferable to the other. Also, the size of the direct grammars is a few times smaller than even the OT footed grammar, so Karttunen OT grammars are certainly not preferable by succinctness. For the OT accounts, a grammar referencing feet is sizeably more succinct than one that references only syllables, showing the utility of feet. It's interesting that the

[13] Although our accounts define the same transduction, that does not mean that the transducers LSmoDirFt, LSmoDirSyl, LSmoMonoFtOT are identical at the machine-level. While any finite-state acceptor can be determinized and minimized to a unique, canonical acceptor [21, Sect. 4.4], the same is not true for finite-state transducers. First, not all finite-state transducers are determinizable [40, p. 587]. Second, minimization of a finite-state transducer does not in general result in a unique transducer [34, p. 29].

direct footed account wasn't notably more succinct than the direct syllable one; this could be because of the narrowness of the scope of phonological phenomena considered here. For instance, patterns of stress shift in Samoan upon affixation can be generalized on the basis of constituents [47], but here we considered only monomorphs. The more phonological processes that reference constituents in the grammar, the more the savings from those constituents.

One thing to stress about the foot-based grammars, is that although they place boundaries (parentheses) in the string language, they are very different from SPE-style "boundary symbol theory" [8, 42]. In our grammars, the use of boundary symbols is not arbitrary; rather a left parentheses signals entering into a sequence of states representing a constituent, and a right parentheses is invariably placed when that sequence of states is completed. As [42] points out, compared to a grammar which references nested units in the prosodic hierarchy, grammars with boundary symbols may be expressive enough to fit the data, but the lack of a well-defined relation between the different boundary symbols makes the grammars much too expressive. Moreover, boundary symbol theory locate the boundary symbols in the alphabet and allows their placement to be restricted only by general restrictions on possible rewrite rules. But we are coding constituency into the state: the LSmo foot-based grammars place parentheses in the string language so we can refer to the units that they enclose, and what restricts their placement is the phonological generalizations defined in the grammar.

Comparing the direct grammars to the OT grammars, a number of the transducers defined are identical or similar, e.g., EdgemostR appears in both footed accounts. This suggests that structural regularities we notice in phonological patterns can be well-described in both types of grammars. However, there is a striking difference between the direct grammars and the OT grammars: the OT grammars have scaling problems. The two direct accounts defined can handle syllable strings of arbitrary length. But for the OT accounts, as the syllable string gets longer, the amount of counting that needs to be done increases. In the foot-based OT account, we defined ParseSyll constraint family only up to 5 syllables. Add another syllable to the syllable string, and the grammar becomes inadequate. In the syllable-based OT account, the NokClash and Align constraint families also effectively count up violations, resulting in the same kind of scaling problem. As previously mentioned, it is in fact the unlimited violation counting that "pushes [standard] optimality theory out of the finite domain" [27, p. 11]. The adequacy of the direct accounts—which are similar in size to the OT grammars and even describe similar regularities—suggests that perhaps this additional power is unnecessary.

It is also an advantage that finite state transducers are sufficient to define the direct grammars. Defining phonology with finite state tranducers is not only helpful as a common formalism with comparison of syntax (and other patterns), but also enables us to compose phonological transducers with syntactic ones to model the syntax-phonology interface. In contrast, the expressive power of finite state transducers is not enough to define OT grammars where underlying forms are mapped directly to surface forms rather than violation vectors. As noted by [27], any constraint that can be multiply violated such as Parse-σ cannot be defined with a finite state transducer in [27]'s formalization of OT because a finite system cannot distinguish between infinitely many degrees of well-formedness [10, 31]. If, instead we do define relations that map from underlying forms to violations as in standard OT, we can easily use xfst to implement Parse-σ as: define ParseSyllOT ["Unparsed" -> "1",\"Unparsed" -> "0"]; [11] The FST is trivial: a 2-state machine, where one state maps any syllable that is not unparsed to 0 and the other maps an unparsed syllable X[?] to 1, e.g., it maps X[L]P[L]X[L] to 101. This shows an advantage to mapping to violations in Con, but then the scaling problem with counting is shifted to Eval. In Karttunen's formalization, both gradient and categorical constraints that can assign multiple violations cannot be defined with a FST. Moreover, as we saw with Align(x1;L,x0,PWd), the implementation of gradient constraints is even more cumbersome. Not only are the machines that xfst compiles them into much too big and redundant to discover generalizations in; even the high-level language description are, too. When OT transduces instead to violation marks, only gradient constraints cannot be defined with a FST. While [31] argues that OT constraints are categorical, even if that is the case, Eval isn't a finite state process. OT isn't regular if the number of violations is unbounded [10].

However, one potential advantage of OT grammars vs. direct grammars in the work here is that limiting ourselves to mainstream constraints proposed by phonologists has provided some restrictions on the defined grammars—even if these restrictions might not be well-characterized in terms of structural classes in the Chomsky or sub-regular hierarchy, and even if standard OT admits supra-regular transductions. Limiting the statement of restrictions in direct grammars to the expressivity of regular transductions (as we've done here) is not restrictive enough: it is well-known that regular transductions include patterns that we'd never expect to see in phonology, e.g., [19]. On-going work has been tightening bounds on the expressivity of 'direct' transductions to sub-regular classes and making connections between restrictions on these transductions and restrictions on OT constraints, see [20] for an overview.

4 Conclusion

In this paper, we implemented and compared syllable- and foot-based grammars of Samoan stress patterns. We made this comparison in Karttunen's finite state formalization of OT, and in grammars directly describing restrictions of the surface patterns. The definition and comparison of the grammars was done in the xfst language to follow linguistic practice, since xfst was designed to be a high-level language that makes it easy to express and detect linguistic generalizations. Such generalizations might not be revealed at the level of a regular grammar or finite state machine.[14] In the OT formalism, having the prosodic constituents of

[14] See the github repository for graphs of the transducers defined for each of the four accounts.

feet clearly allowed the grammar to be much more succinct. However, whether or not we have feet in the direct account did not impact succinctness of the grammar. It is striking that direct finite-state descriptions of phonological patterns have revealed strong structural universals without referring to constituents, while the advantage of having constituents is clear in the OT formalism used here. The difference in the comparison between the two types of grammars may simply be because our measure of succinctness is not appropriate, and also may not hold in general, or because the range of phonological phenomena considered here is too narrow.

A natural follow-up to the work here would be to extend grammar comparisons to a wider range of phonological phenomena that have been studied in prosodic phonology. For instance, all the dependencies in the Little Samoan language defined here are local. What if the language included non-local dependencies? Another natural follow-up would be to explore the consequences of introducing constituents in more expressive grammars. For instance, OpenFST is a finite state transducer library that offers the capability to define grammars with the expressivity of context-free languages via pushdown automata, which are finite state transducers augmented with a stack [1,2]. It would be interesting to see if a comparison of syllable-based and foot-based grammars for Samoan stress defined with pushdown automata might yield different results from the ones here. More broadly, this paper shows a way in which we can study concrete, specific linguistic proposals and engage closely with linguistic practice, while still maintaining a rigorous approach. We hope that this proof of concept may inspire additional computational work taking this kind of approach.

References

1. Allauzen, C., Riley, M.: A pushdown transducer extension for the OpenFst library. In: Moreira, N., Reis, R. (eds.) CIAA 2012. LNCS, vol. 7381, pp. 66–77. Springer, Heidelberg (2012). https://doi.org/10.1007/978-3-642-31606-7_6
2. Allauzen, C., Riley, M., Schalkwyk, J., Skut, W., Mohri, M.: OpenFst: A general and efficient weighted finite-state transducer library. In: Holub, J., Žd'árek, J. (eds.) CIAA 2007. LNCS, vol. 4783, pp. 11–23. Springer, Heidelberg (2007). https://doi.org/10.1007/978-3-540-76336-9_3
3. Bailey, T.: Nonmetrical constraints on stress. Ph.D. thesis, University of Minnesota (1995)
4. Beesley, K.R., Karttunen, L.: Finite State Morphology. CSLI, Stanford, CA (2003)
5. Berwick, R.C.: Mind the gap. In: Gallego, A., Ott, D. (eds.) 50 Years Later, MITWPL77, pp. 1–12. MIT, Cambridge, Massachusetts (2015)
6. Bird, S., Ellison, T.M.: One level phonology: autosegmental representations and rules as finite automata. Comput. Linguis. **20**, 55–90 (1994)
7. Chomsky, N.: Three descriptions of language. IRE Trans. Inf. Theory **2**(3), 113–124 (1956)
8. Chomsky, N., Halle, M.: The Sound Pattern of English. The MIT Press, Cambridge (1968)
9. Eisner, J.: Efficient generation in primitive optimality theory. In: Proceedings of the 35th Annual Meeting of the Association for Computational Linguistics (1997)

10. Frank, R., Satta, G.: Optimality theory and the generative complexity of constraint violability. Comput. Linguist. **24**(2), 307–315 (1998)
11. Gainor, B., Lai, R., Heinz, J.: Computational characterizations of vowel harmony patterns and pathologies. In: Choi, J., et al. (eds.) WCCFL 29, pp. 63–71. Cascadilla Proceedings Project, Somerville, MA (2012)
12. Goldsmith, J.A.: Autosegmental phonology. Ph.D. thesis, MIT (1976)
13. Gordon, M.: A factorial typology of quantity insensitive stress. NLLT **20**, 491–552 (2002)
14. Hartmanis, J.: On the succinctness of different representations of languages. SIAM J. Comput. **9**, 114–120 (1980)
15. Hayes, B.: Metrical Stress Theory. University of Chicago Press, Chicago (1995)
16. Hayes, B., Tesar, B., Zuraw, K.: Otsoft 2.4. Software package (2016). www.linguistics.ucla.edu/people/hayes/otsoft/
17. Heinz, J.: On the role of locality in learning stress patterns. Phonology **26**(02), 303–351 (2009)
18. Heinz, J.: Learning long-distance phonotactics. LI **41**(4), 623–661 (2010)
19. Heinz, J.: Computational phonology - part I: foundations. Lang. Linguist. Compass **5**(4), 140–152 (2011)
20. Heinz, J.: The computational nature of phonological generalizations. In: Hyman, L.M., Plank, F. (eds.) Phonological typology. De Gruyter Mouton, Berlin/Boston (2018)
21. Hopcroft, J.E., Motwani, R., Ullman, J.D.: Introduction to Automata Theory, Languages, and Computation. Pearson Education, Boston (2007)
22. Hulden, M.: Finite state syllabification. In: Yli-Jyrä, A., Karttunen, L., Karhumäki, J. (eds.) FSMNLP 2005, pp. 120–131. Springer, Berlin (2006)
23. Hyde, B.: Non-finality and weight sensitivity. Phonology **24**(2), 287–334 (2007)
24. Johnson, C.D.: Formal aspects of phonological description. Mouton (1972)
25. Kager, R.: Rhythmic licensing theory: an extended typology. In: Proceedings of the 3rd Seoul International Conference on Phonology, pp. 5–31 (2005)
26. Kaplan, R.M., Kay, M.: Regular models of phonological rule systems. Comput. Linguist. **20**(3), 331–378 (1994)
27. Karttunen, L.: The proper treatment of optimality in computational phonology. In: FSMNLP 1998 (1998)
28. Kenstowicz, M.: Cyclic vs. non-cyclic constraint evaluation. Phonology **12**, 397–436 (1995)
29. Kiparsky, P.: Word formation and the lexicon. In: Ingemann, F. (ed.) Proceedings of the 1982 Mid-America Linguistics Conference, pp. 3–29. University of Kansas, Lawrence (1982)
30. Kornai, A.: Formal phonology. Ph.D. thesis, Stanford University (1991)
31. McCarthy, J.J.: OT constraints are categorical. Phonology **20**(1), 75–138 (2003)
32. McCarthy, J.J., Prince, A.S.: Generalized alignment. In: Booij, G., Van Marle, J. (eds.) Yearbook of Morphology, pp. 79–153. Springer, Dordrecht (1993). https://doi.org/10.1007/978-94-017-3712-8_4
33. Meyer, A., Fischer, M.: Economy of description by automata, grammars, and formal systems. In: SWAT 1971, pp. 188–191 (1971)
34. Mohri, M.: Finite-state transducers in language and speech processing. Comput. Linguist. **23**, 1–42 (1997)
35. Nespor, M., Vogel, I.: Prosodic Phonology. Foris Publications, Dordrecht (1986)
36. Prince, A., Smolensky, P.: Optimality theory: Constraint interaction in generative grammar. ROA version, 8/2002, Rutgers University Center for Cognitive Science (1993)

37. Prince, A., Smolensky, P.: Optimality Theory: Constraint Interaction in Generative Grammar. Blackwell Publishing, Malden (2004)
38. Rasin, E., Katzir, R.: On evaluation metrics in Optimality Theory. LI (To appear)
39. Rissanen, J.: Stochastic Complexity in Statistical Inquiry Theory. World Scientific Publishing Co. Inc., River Edge (1989)
40. Sakarovitch, J.: Elements of Automata Theory. Cambridge University Press, Cambridge (2009). Translated by Reuben Thomas edn.
41. Salomaa, A.: Formal Languages. Academic Press, New York (1973)
42. Selkirk, E.: Prosodic domains in phonology: Sanskrit revisited. In: Aronoff, M., Keans, M.L. (eds.) Juncture. Anma Libri, Saratoga, California (1980)
43. Selkirk, E.O.: Phonology and Syntax. MIT Press, Cambridge (1986)
44. Stabler, E.P.: Two models of minimalist, incremental syntactic analysis. Top. Cogn. Sci. **5**(3), 611–633 (2013)
45. Wagner, M., Watson, D.G.: Experimental and theoretical advances in prosody: a review. Lang. Cogn. Processes **25**, 905–945 (2010)
46. Zuraw, K.: Prosodic domains for segmental processes?, June 2009. https://www.mcgill.ca/linguistics/files/linguistics/Handout_RevisedForMcGill.pdf
47. Zuraw, K., Yu, K.M., Orfitelli, R.: The word-level prosody of Samoan. Phonology **31**(2), 271–327 (2014)

Modelling Derivational Morphology: A Case of Prefix Stacking in Russian

Yulia Zinova(✉)

Heinrich Heine University, Düsseldorf, Germany
zinova@phil.hhu.de

Abstract. In order to automatically analyse Russian texts, one needs to model complex verb formation, as it is a productive mechanism and dictionary data is not sufficient. In this paper I discuss two implementations that aim to produce all and only the existing complex verbs built from the available morpheme inventory for the same fragment of Russian grammar. The first implementation is based on the syntactic theory approach to prefix combinatorics by Tatevosov (2009) and the other one uses the combination of basic syntactic restrictions and frame semantics to construct all possible combinations. I show that a combination of basic syntactic and semantic restrictions provides better results than a set of elaborated syntactic restrictions, especially for the complex verbs that are not normally tested by introspection.

1 Russian Verbal Prefixation System

Russian verbal derivational morphology is extremely rich. One stem can serve as a base for deriving hundreds of verbs via prefixation and suffixation. This is possible due to a high number of prefixes (28 according to Švedova 1982, p. 353, whereby most of them have productive usages), polysemy of prefixes (e.g., the prefix *pere-* has 10 usages according to Švedova 1982, pp. 363–364), and the possibility of prefix stacking. In addition to this, at some stages of the derivation (once per derivation) the imperfective suffix can be attached to the verb. The number and order of affixes, in turn, influences the aspect: prefixation usually leads to the perfective aspect of the derived verb and suffixation leads to the imperfective aspect of the derived verb. On the other hand, both prefixation and suffixation processes are restricted: not any prefix can be attached to a given verb (either simplex or complex), and the suffixation is not always available.

To show how the whole system functions together, let me provide an example. We start with a simplex verb *pisat'* 'to write'. It is imperfective and refers to an unbounded writing activity. If it is prefixed with *za-*, the derived verb *zapisat'* 'to record' is perfective and refers to a completed event of recording something. It can be, in turn, suffixed, and the derived verb *zapisyvat'* 'to record/be recording' is imperfective. Yet another prefixation step can be made, for example with the prefix *do-*, which results in the derived perfective verb *dozapisyvat'* 'to finish recording again', as shown in (1).[1]

[1] IPF superscript marks the imperfective aspect of the verb and PF superscript marks the perfective aspect of the verb.

A. Foret et al. (Eds.): FG 2017, LNCS 10686, pp. 125–141, 2018.
https://doi.org/10.1007/978-3-662-56343-4_8

(1) pisat'^{IPF} → zapisat'^{PF} → zapisyvat'^{IPF} → dozapisyvat'^{PF}
'to write' → 'to record' → 'to (be) record(ing)' → 'to finish recording'

On the other hand, the order of the last two steps can be reversed: if the prefix *do-* is attached to the verb *zapisat'* 'to record', the derived verb *dozapisat'* 'to finish recording' is perfective and can be suffixed with *-iva-*, producing the imperfective verb *dozapisyvat'* 'to finish/be finishing recording', as shown in (2). As a result, the verb *dozapisyvat'* 'to finish/be finishing recording' can be obtained through two different derivations, one leading to the perfective (1) and the other (2) leading to the imperfective aspect of the derived verb.

(2) pisat'^{IPF} → zapisat'^{PF} → dozapisat'^{PF} →
'to write' → 'to record' → 'to finish recording' →
dozapisyvat'^{IPF}
'to be finishing recording'

To illustrate the limits of the derivational morphology, let us try to change the order of prefix attachment in the derivation (2). If the prefix *do-* is attached first, the derived verb *dopisat'* 'to finish writing' exists and is perfective. It is, however, not possible to attach the prefix *za-* to it: the verb **zadopisat'* does not exist (3).

(3) pisat'^{IPF} → dopisat'^{PF} *→ *zadopisat'
'to write' → 'to finish writing'

The goal of both accounts I discuss in this paper is to predict which complex verbs can be derived using the given set of morphemes and which aspect they will have. The implemented grammar fragment contains the following elements: a verb *pisat'* 'to write', a prefixed verb *zapisat'* 'to record', prefixes *po-* (delimitative and distributive interpretations), *pere-* (repetitive and distributive interpretations), and *do-* (completive interpretation), and the imperfective suffix *-iva-*. With this inventory I construct verbs with a maximum of four affixes (this can be realised if the base verb is prefixed two times, then suffixed, and then prefixed again).

Two alternative implementations proposed in this paper are based on two approaches to the prefixation system. In Sect. 2, I present the syntactic approach which is used as a base for the first implementation. Then, in Sect. 3, I introduce the frame semantic approach which motivates the second implementation. I then show that replacing complex syntactic restrictions with a combination of simple syntactic restrictions and semantic restrictions allows for better predictions of the existence and aspect of complex verbs with respect to both precision and recall.

2 Syntactic Approach

2.1 Theory

The main idea of approaches that pursue a syntactic view of the prefixation is to represent the verbal structure by means of a syntactic tree (Babko-Malay 1999) and divide various prefix usages into categories such that each category is related to a specific position in the tree. This allows to restrict the available derivations. For example, according to this view, the derivation (3) is blocked because the prefix *za-* cannot occupy the position higher than that of the prefix *do-*, as the prefix *za-*, with the usage it has in the verb *zapisat'* 'to record', is classified as lexical whereas the prefix *do-* is classified as superlexical.

These two categories—lexical and superlexical prefixes—form a base of any existent syntactic approach to prefixation, as the two classes are claimed to exhibit distinct properties due to the different structural positions in the verbal tree. Such an approach is pursued by Svenonius (2004a, b), Romanova (2006), Ramchand (2004), Tatevosov (2007, 2009), among others. A serious problem of all these accounts is that they implicitly predict the non-existence of biaspectual verbs, as the highest affix in the structure serves to determine the aspect of the whole verb and there is only one possible structure for any verb with a fixed interpretation.[2]

In this paper I address the theory proposed by Tatevosov (2009) that stems from this line of research but includes substantial modifications with respect to the earlier proposals. Tatevosov (2009) divides the class of superlexical prefixes into three groups: selectionally limited, positionally limited, and left periphery prefixes.

Selectionally limited prefixes can be added only to a formally imperfective verb. The group includes the delimitative prefix *po-* (*posidet'* 'to sit for some time'), the cumulative prefix *na-*, the distributive prefix *pere-* (*perelovit' X* 'to catch all of X'), and the inchoative prefix *za-*. The group of positionally limited prefixes is constituted by the completive prefix *do-* (*dodelat'* 'to finish doing'), the repetitive prefix *pere-* (*perepisat'* 'to rewrite'), and the attenuative prefix *pod-*. These prefixes, according to Tatevosov (2009), can be added only before the secondary imperfective suffix *-yva-/-iva-*. The group of left periphery prefixes is constituted by only one prefix: distributive *po-* (*pobrosat'* 'to throw all of'). It occupies the left periphery of the verbal structure.

Such a division of superlexical prefixes into several subclasses allows Tatevosov (2009) to effectively limit the number of complex verbs. One drawback of the analysis is, as mentioned above, the prediction of the absence of complex biaspectual verbs (missing verb-aspect pairs). In order to better test the accuracy of the proposal by Tatevosov (2009), I have implemented it for the grammar fragment described above. In the next section I show fragments of the implementation and explain decisions that I had to make.

2.2 Implementation

For the implementation, I have used EXtensible MetaGrammar[3] (XMG, Crabbé et al. 2013; Petitjean et al. 2016)—a formalism that allows to describe linguistic

[2] In any existent syntactic analysis either (1) or (2) is not a valid derivation. For details, see Zinova and Filip (2013).

[3] http://xmg.phil.hhu.de/.

information contained in the grammar and a tool to compute grammar rules and produce a redundant, strongly lexicalised Tree Adjoining Grammar (TAG, Joshi and Schabes 1997). In particular, the compiler for the current implementation is created using XMG 2 and has a syntactic (syn) and a frame semantic (frame) dimension (Lichte and Petitjean 2015). The code and the xml file that is output by the compiler are available online.[4]

The syntactic dimension is described using the following elements: first, all the nodes are declared using the keyword `node` and a variable name. These declarations are accompanied by optional marks (in brackets) and syntactic features (in square brackets, separated by commas). Values of syntactic features can be either specified or represented by a variable to ensure the same value of the feature across the nodes without specifying it. Second, the relations between the nodes are stated (immediate dominance, dominance, immediate precedence, precedence).

XMG is designed to output unanchored TAG elementary trees, but as currently there is no parser that would take into account the frame semantic dimension, I simulate the insertion of lexical anchors in the metagrammar. This solution leads to a more complicated metagrammar architecture, but allows us to see the results in a form that can be easily understood. If I were to output the unanchored trees only, I would obtain prefixation schemes but the stem that carries important information would not be inserted, which would make it very hard to check the predictions of the second implementation. (For the first implementation it would not make much difference as the only property of the verbal stem that can influence prefixation patterns in their productive part is aspect.)

As the first implemented approach is syntactic, all restrictions are formulated in syntactic terms and the frame dimension is used to represent the order of attachment of affixes with different semantics. For example, the class for the distributive interpretation of the prefix *pere-* looks as shown on Fig. 1. The

```
class PereVerb
export ?VP ?VPInt
declare ?VP ?VPInt ?Pere ?PereLex ?AGR ?X0 ?X1
{ <syn>{
    node ?VP [cat=vp, agr=?AGR, e=?X1, aspect = perf];
    node ?Pere [cat=pref];
    node ?PereLex (mark=flex) [cat=pere-];
    node ?VPInt [cat=vp, agr=?AGR, e=?X0, aspect = imperf];
    ?VP -> ?VPInt; ?VP -> ?Pere; ?Pere -> ?PereLex; ?Pere >> ?VPInt
  }; <frame>{
    ?X1[distributive,
      of: ?X0] } }
```

Fig. 1. XMG implementation for the distributive interpretation of the prefix *pere-* following Tatevosov (2009)

[4] https://user.phil-fak.uni-duesseldorf.de/zinova/XMG/index.html.

restriction on this prefix attachment is the imperfective aspect of the base verb, which is reflected via a syntactic constraint on the feature *aspect* of the node `?VPInt`. The top node after the prefixation is `?VP` and it is characterized by the *perf* value of the feature *aspect*. The semantic dimension is a dummy for storing the relevant usage labels (as they are related to distinct syntactic properties) and keeping track of the approximate meaning of the derived verb so it can be compared with the exact meaning we will be dealing with in the second interpretation.

Other constraints (restricting the position of the prefix either to that below the imperfective suffix or to the leftmost slot in the structure) are realized through limiting the classes that can be assembled with the derivational base at each step.

The output of this implementation consists of 81 models. Each model is associated with a certain verb with fixed order and interpretation of the prefixes. As an example, let me show the output for the verb *dozapisyvat'* 'to (be) finish(ing) recording' that we have discussed in Sect. 1. On Fig. 2, one can see that the last step of the derivation is the attachment of the imperfective suffix and the aspect of the derived verb is imperfective. This corresponds to the derivation (2). There is no other model in the output that would correspond to the derivation (1).

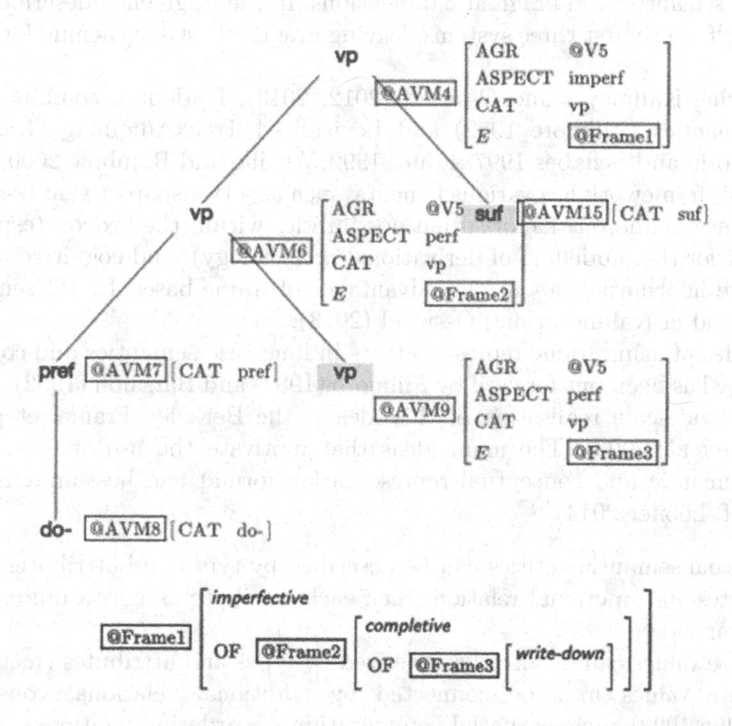

Fig. 2. XMG model for the verb *dozapisyvat'* 'to (be) finish(ing) recording'

To evaluate the output, I have manually checked the existence of all the possible verb-aspect pairs using the assumption that an existent verb cannot be derived from a non-existent base. The total number of possible complex verbs built from the given morpheme inventory and paired with aspect is 546. The number of existent verb-aspect pairs that can be created out of this set, according to my count, is 70. Out of them the implementation of the account proposed by Tatevosov (2009) produces 52, which amounts to 0,642 precision and 0,743 recall for the implemented grammar fragment. In the next section I show how this result can be improved if complex syntactic restrictions are replaced with a combination of basic syntactic restrictions with semantic restrictions.

3 Alternative: Frame Semantics

3.1 Framework

I argue that Russian verbal prefixation is a complex system that cannot be successfully modelled by means of one linguistic layer. In order to simplify individual components of the system and allow for the observed flexibility without massive overgeneration, one needs to coordinate the work of the morphological, syntactic, semantic, and pragmatic dimensions. In the fragment I describe here I limit myself to the first three systems, leaving pragmatic strengthening for future work.

Following Kallmeyer and Osswald (2012, 2013), I adopt a combination of frame semantics (Fillmore 1982) and Lexicalized Tree Adjoining Grammars (LTAG, Joshi and Schabes 1997; Frank 1992; Abeillé and Rambow 2000; Frank 2002). This framework has various benefits, such as a transparent syntax-semantics interface, numerous factorisation possibilities within the lexicon (especially important for the modelling of derivational morphology), and cognitive motivation. More information about the advantages of frame-based LTAG semantics can be found in Kallmeyer and Osswald (2013).

The idea of using frame representations in linguistic semantics and cognitive psychology has been put forward by Fillmore (1982) and Barsalou (1992), among others. A widescale realisation of this idea is the Berkeley FrameNet project (Fillmore et al. 2003). The main ideas that motivate the use of frames as a general semantic and conceptual representation format can be summarized as follows (cf. Löbner 2014):

- conceptual-semantic entities can be described by types and attributes;
- attributes are functional relations, i.e., each attribute assigns a unique value to its carrier;
- attribute values can be also characterized by types and attributes (recursion);
- attribute values may be connected by additional relational constraints (Barsalou 1992) such as spatial configurations or ordering relations.

These ideas are formalized by Kallmeyer and Osswald (2013) who define frames as finite relational structures in which attributes correspond to functional relations. The members of the underlying set are referred to as the *nodes* of the frame. An important restriction is that any frame must have a *functional backbone.* This means that every node has to be accessible via attributes from at least one of the *base nodes*: nodes that carry *base labels* (unique identifiers). Importantly, feature structures may have multiple base nodes. In such a case often some nodes that are accessible from different base nodes are connected by a relation.

Another important component of the formalism is the type hierarchy. Since the number of syntactic restrictions I use is very limited, many derivations will be filtered out by the semantic constraints. For this, there are two main mechanisms: unification failure (type incompatibility or conflicting attribute values) and constraint failure (requirement for the two values to be in a specific relation is not satisfied). It is important to note that in the formalisation of Frame Semantics proposed by Kallmeyer and Osswald (2013) all types are considered compatible unless stated otherwise, so all the type conflicts have to be listed explicitly.

3.2 Frame Representations for Selected Prefixes

In this section I show frame representations of prefixes that are included in the implemented grammar fragment: *po-, pere-, do-*. Due to the space constraints I present only the representations and skip the theoretical motivation for them.

The Prefix *po-*. The first prefix I provide a frame representation for is *po-*. The usages that are of interest for the implementation are the delimitative (*posidet'* 'to sit for some time') and the distributive (*pobrosat'* 'to throw all of') ones. On the basis of the discussions in (Filip 2000; Kagan 2015; Zinova 2017, Chap. 4) I propose to represent the contribution of the prefix by the frame shown on the left side of Fig. 3. The idea behind this representation is that the prefix adds information that the event is bounded (type *bounded-event*) and the initial (INIT) and the final (FIN) stages of the event are related to arbitrary points on the scale (2 and 3 are free variables).

Such a representation allows the derivation of both delimitative and distributive usages of the prefix when the appropriate scale is selected. The equivalence of VERB-DIM and M-DIM attributes means that the appropriate scale must be equivalent to the verbal dimension. For the verb *pisat'* 'to write' this would be the time dimension that is realised as self-scaling. When this dimension is selected, the delimitative interpretation is acquired. If there is a source of iteration (e.g., a quantified object), the type of the M-DIM gets conjuncted with *cardinality* and the derived verb is interpreted distributively.

The Prefix *pere-*. The prefix *pere-* is extremely polysemous. In this paper I consider two of its usages that are relevant for the implemented grammar fragment: distributive (*perelovit' X* 'to catch all of X') and repetitive (*perepisat'* 'to rewrite').

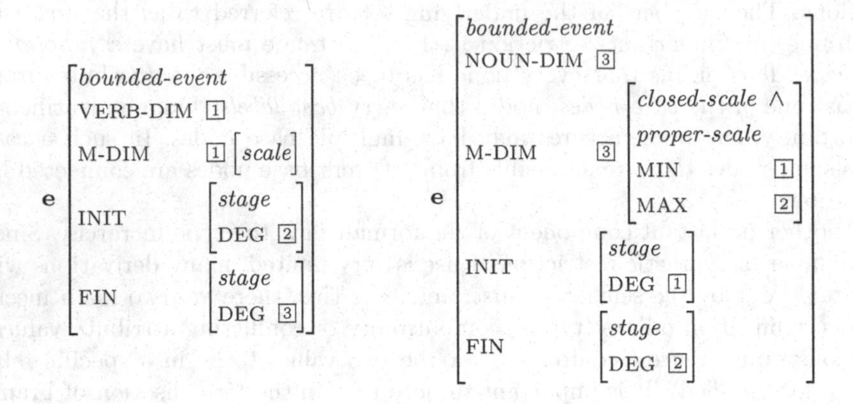

Fig. 3. Frame representations of the prefixes *po-* (left) and *pere-* (case of a closed scale, right) and frame for coercion of an unbounded event into a bounded event (right)

The frame for the distributive usage of the prefix *pere-* is shown on the right side of Fig. 3. The key restrictive factor in this case is the type of the measure dimension (M-DIM) that has to be supplied by the context (M-DIM = NOUN-DIM, the context-determined dimension is called NOUN-DIM as usually it is the direct object that supplies it): a closed proper scale in this case (*closed-scale* ∧ *proper-scale*). The initial and final stages of the event correspond to the minimum and maximum points on the scale (INIT.DEG = M-DIM.MIN; FIN.DEG = M-DIM.MAX).

The repetitive usage of the prefix *pere-* arises when the measure dimension of the event denoted by the derivational base (PREP.M-DIM) is of type *property-scale* (Fig. 4, the types *proper-scale* and *property-scale* are not compatible with each other). This event then becomes a value of the preparatory phase (PREP) attribute of the new event. The initial and the final stages, the noun dimension, the measure dimension, and the manner attributes are copied to the event node that refers to the new event (M-DIM = PREP.M-DIM, FIN = PREP.FIN, INIT = PREP.INIT, MANNER = PREP.MANNER). The tree on the right side of 4 shows that the derived verb refers to a frame node (feature E, E = **f**) other than the derivational base (E = **e**). It also stores information about the noun to the right of the verb being the THEME of the original event (PREP.THEME = 6, I = 6).

The next restriction for the repetitive usage of the prefix *pere-*, apart from the property type of the scale, is that the event denoted by the derivational base must have a final stage in its representation. This means that a simplex imperfective verb cannot be combined with this prefix usage, unless it is coerced into a bounded event. On the formal side it means formulating a requirement on the frame configuration (the presence of the FIN attribute). For implementing the coercion of an unbounded event into a bounded event, I propose to use the frame

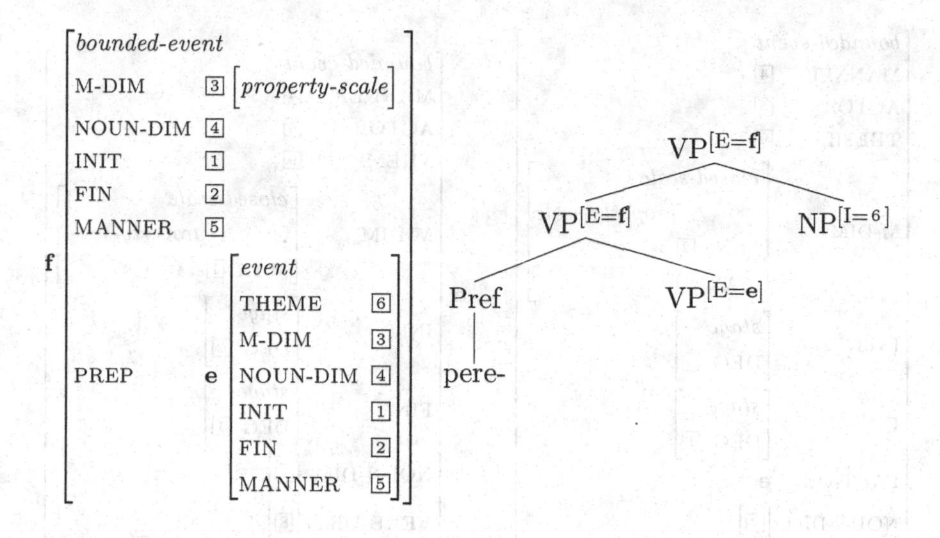

Fig. 4. Representation of the contribution of the prefix *pere-*: case of a property scale

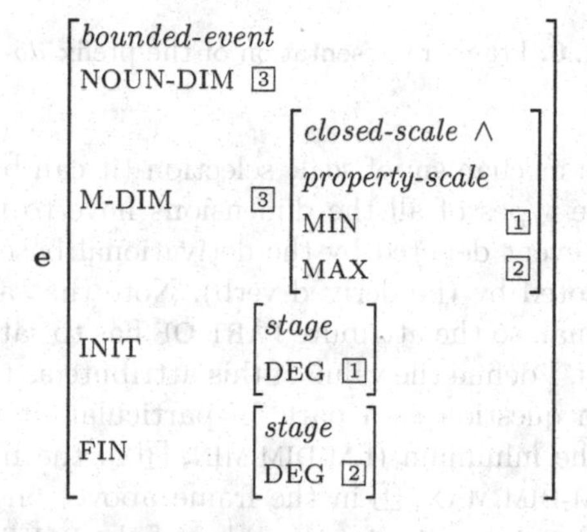

Fig. 5. Frame for coercion of an unbounded event into a bounded event

shown on Fig. 5. It functions similarly to the frame for the distributive usage of the prefix *pere-* (on the right side of Fig. 3), varying from it only with respect to the type of the measure dimension (*property-scale* instead of *proper-scale*).

The Prefix *do-*. The last prefix included in the implemented grammar fragment is the prefix *do-* with completive semantics. The event denoted by a *do*-prefixed verb is a terminal part of the original event. In other words, when the prefix is attached, the maximum of the scale has to be associated with the final stage of the event (M-DIM.MAX = FIN.DEG). The frame shown on Fig. 6 realizes this as

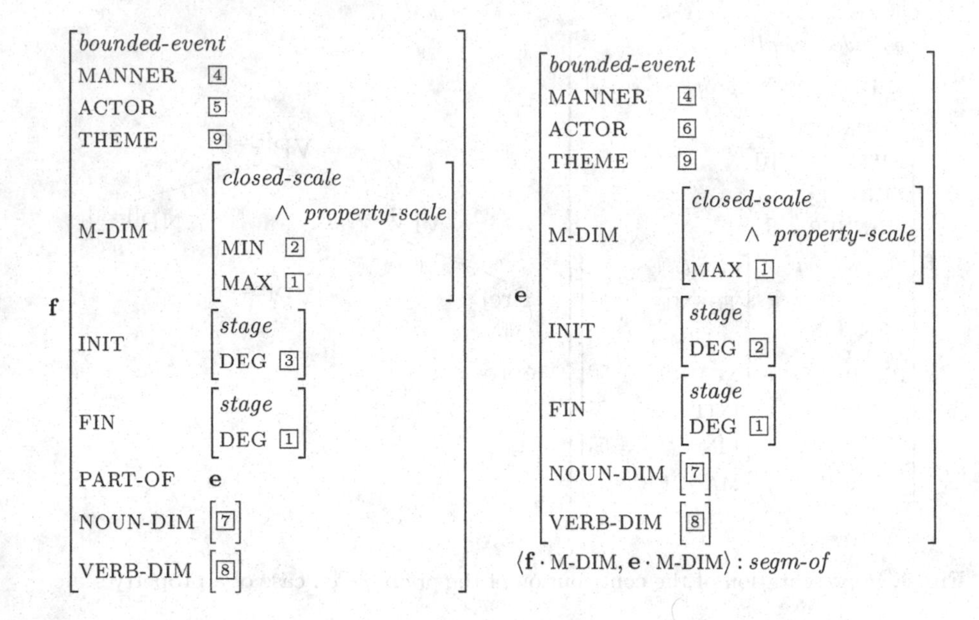

Fig. 6. Frame representation of the prefix *do-*

well as the *do*-specific mechanism of scale selection (it can be either NOUN-DIM or VERB-DIM, but the types of all the dimensions have to be copied from the representation of the event denoted by the derivational base to the representation of the event denoted by the derived verb). Note that attributes in Frame Semantics are functional, so the attribute PART-OF has to satisfy this restriction as well. To ensure this, I define the value of this attribute as the maximum event that has the event in question as a part. In particular, it would be an event that proceeds from the minimum (**f**.M-DIM.MIN, 3 in the frame above) to the maximum degree (**f**.M-DIM.MAX, 1 in the frame above) on the relevant scale. The scale has to be closed in order for the value of the PART-OF attribute to be defined using the frame on the left side of Fig. 6.

Similarly to the iterative usage of the prefix *pere-*, the prefix *do-* can be only attached to bounded events. This means that, again, simplex imperfective verbs need to be first coerced into a bounded interpretation.

3.3 Implementation

Restrictions. In this section I present the implementation of the frame-based proposal for limiting the derivation of complex verbs and predicting their aspect and semantics. As the frame domain is a new development in XMG, I had to deal with parser restrictions and place some semantically motivated constraints on the syntactic level.

First, I had to 'lift' two features (*bounded* and *limited*) to the syntactic level due to the fact that such feature checking inside the semantic dimension of XMG is not yet supported. The feature *bounded* appears at those nodes that are associated with frames of *event* type. It gets the value *yes* if there is a path from the central node of the frame to an attribute FIN that can proceed through the PART-OF attributes. This corresponds to the event being of type *bounded-event* or being a PART-OF an event of a *bounded-event* type. If there is no such path, the value of the feature is *no.* The feature *limited* is a stronger version of a similar constraint: for *limited* to get the value *yes*, the central node has to have an attribute FIN and its value has to be specific (concrete value or a bound variable). This corresponds to the event being of type *bounded-event.* In all other cases the feature *limited* gets the value *no.*

Another restriction that is located in the syntactic domain despite its semantic nature is the *unicity of iteration.* The idea is that inside a derivation there can be only one semantic marker of iteration. In the current implementation, this feature is doubled on the syntactic side because there is no possibility to check whether this constraint holds on the semantic side.

Type Hierarchy. As I have noted above, type incompatibility is one of the mechanisms that blocks derivations. The type hierarchy description consists of two types of statements: (1) statements declaring that one type is also some other type, e.g. `property-scale -> scale` (something of type *property-scale* is also of type *scale*) and (2) statements declaring that types are not compatible, e.g. `cardinality property-scale -> -` (something of type *cardinality* cannot simultaneously be of type *property-scale*). In the implementation proposed here I postulate a minimum set of constraints that is sufficient to block unwanted prefix combinations and is motivated by the internal structure of the scales in question.

Lexical Anchors. In a proper implementation that would separate the metagrammar level from the syntactic level the following elements would not belong to the metagrammar, but would be used as lexical anchors for the appropriate tree families. The first entry is the noun that is used to fill the object slot. For the implementation proposed here I have selected the plural form of the noun *rasskaz* 'story'. For our purposes it is important that stories have length and (due to the plurality of the noun) also some cardinality.

The description of the noun (Fig. 7) is straightforward: on the syntactic side, it is a daughter of the N category node and on the semantic side it contains relevant attributes. The two nodes (?N and ?Story) are declared in the first two lines of the syntactic domain description and connected via an immediate dominance relation in the third line.[5] Both nodes are characterized with feature

[5] This unary branching is necessary in the current implementation due to the absence of a syntactic compiler that would work with frames: it models the lexical anchor insertion inside the metagrammar.

```
class Story
export ?Length ?Card ?N
declare ?N ?Story ?X0 ?Length ?Card
{ <syn>{
    node ?N (mark=coanchor) [cat=n, num = pl, i=?X0];
    node ?Story (mark=nounacc) [cat=rasskazy, num = pl, i=?X0];
    ?N -> ?Story
  }; <frame>{
    ?X0[story,
        length: ?Length,
        cardinality: ?Card ] } }
```

Fig. 7. XMG code: noun that is used to fill the accusative NP slot

i=?X0 which connects them to the semantic frame characterized in the frame dimension. The frame description states that the type of the frame ?X0 is `story` and it has two attributes: length and cardinality.

These attributes enable the noun to enter one of the dimension constructors. The cardinality constructor is available for all nouns that have a cardinality attribute with an additional restriction for plural number. On the semantic side it creates an M-DIM attribute and the event description bounded to the VP node also acquires the type *iteration*. Another dimension constructor that I use, implemented in the class NounLength, creates a NOUN-DIM of type *property-scale* ∧ *length* and the MAX being equal to the value of the LENGTH attribute of the noun.

The second group of lexical items consists of two verbs: *pisat'* 'to write' and *zapisat'* 'to write down'. The second verb contains the prefix *za-*, but its semantic contribution is not transparent (in terms of a syntactic approach it is a lexical prefix usage), so the whole verb must be stored in the dictionary. The class that represents the verb *pisat'* 'to write' has a simple syntactic structure of two nodes (see Fig. 8): the node of category V and the node that contains the verb itself, where the V node inherits all syntactic properties of the verb, except for the category. The *aspect* feature, in contrast with the features *limited* and *bounded*, is a syntactic feature and carries information about the syntactic aspect of the verb represented by the respective node. For the frame semantic side, I use a simple representation that serves the purposes of the current analysis.

Prefixes. As we have already discussed the frames for all individual prefix usages in Sect. 3.2, I will now show how frames correspond to the XMG descriptions and what happens on the syntactic side, taking one prefix as an example.

Figure 9 shows the code for the prefix *po-*. The syntactic part of it represents a VP that consists of a prefix head and another (internal) VP that carries information about the derivational base. The agreement information as well as the semantic frame are then passed to the higher VP node. This node is also characterized by the `perf` (perfective) value of the aspect feature independently

```
class Pisat
export ?V
declare ?V ?Pisat ?X0 ?Actor ?Theme ?Mean
{ <syn>{
    node ?V (mark=anchor) [cat=v, e=?X0, asp = unbound, aspect = imperf];
    node ?Pisat (mark=flex) [cat=pisat, e=?X0, asp = unbound,
     aspect = imperf];
    ?V -> ?Pisat
  }; <frame>{
    ?X0[event & process,
       actor:?Actor,
       theme:?Theme,
       mean:?Mean,
       manner:[write],
       verb-dim:?X0 ] } }
```

Fig. 8. XMG code for representation of the verb *pisat'* 'to write'

```
class PoVerb
export ?VP ?VPInt
declare ?VP ?VPInt ?Po ?PoLex ?AGR ?X0 ?Init ?Fin ?VDim
{ <syn>{
    node ?VP [cat=vp, agr=?AGR, e=?X0, limited = yes, bounded = no,
      aspect = perf];
    node ?Po [cat=pref];
    node ?PoLex (mark=flex) [cat=po-];
    node ?VPInt [cat=vp, agr=?AGR, e=?X0, bounded = no];
    ?VP -> ?VPInt; ?VP -> ?Po; ?Po -> ?PoLex; ?Po >> ?VPInt
  }; <frame>{
    ?X0[bounded-event,
       m-dim: ?VDim,
       verb-dim: ?VDim,
       init: [stage,
           scale-deg:?Init],
       fin: [stage,
           scale-deg:?Fin] ] } }
```

Fig. 9. XMG code for the class describing the prefix *po-*

of the value of the aspect feature of the internal VP node. Following the definitions provided above, the feature *limited* is assigned the value *yes* because the semantic frame contains the attribute FIN, but the feature *bounded* is assigned the value *no*, as the value of the attribute FIN is a free variable.

As for the frame description part, it follows the proposed frame configuration straightforwardly. This is evident if one compares the code with the frame on the left side of Fig. 3 for the prefix *po-*.

Encoding of other prefix usages proceeds in a similar way: the syntactic part does not vary much from prefix to prefix and semantic descriptions can be directly obtained from the frame descriptions I have proposed in Sect. 3.2. The coersion that is sometimes required before the attachment of the prefix *pere-* or the prefix *do-* is realised by the class NDimCoercedVerb that transforms a non-bounded event into a bounded event using the nominal scale.

Imperfective Suffix. I use two separate classes to produce two interpretations of secondary imperfective verbs: progressive and habitual. For the analysis that I propose it is important to distinguish between them when another prefix is attached after the suffixation, as these two interpretations have different semantic properties.

The habitual interpretation of the imperfective suffix (*IterVerb* class) produces an unlimited event that is a series of limited events. The NOUN-DIM of the new event necessarily is of type *cardinality* and does not need to correspond to the respective attribute of the derivational base. The verbal dimension is copied from the individual event level to the series level. This interpretation of the imperfective suffix is also associated with the introduction of the *iteration* type of the event and the respective syntactic feature.

The second interpretation of the imperfective suffix is progressive: on the semantic side I represent it as a creation of a new event that is a PART-OF the event denoted by the derivational base. Due to the PART-OF relation the new event remains limited. As relations are currently not implemented in XMG, for the sake of the implementation I use PART-OF as an attribute when representing the progressive interpretation of the imperfective suffix (class *ProgrVerb*), although it is not functional.

Assembling the Parts. The last part of the code assembles the verbal phrases from the components described above. This is done by the classes *OneBasePrefixedVerb*, *VerbWithOnePrefix*, *TwoPrefixedVerb*, *TwoPrefixedSuffixedVerb*.[6]

The compilation of the code produces 88 models (in this case not verbs, but verbal phrases). As these models include two interpretations of the imperfective suffix whereas such distinction is not made in the analysis of Tatevosov (2009), before calculating precision and recall one has to remove 'duplicates': models that differ only with respect to suffix interpretations. This leaves 79 models of which 70 are correct and amounts to 0,886 precision and full recall. Nine extra models have to be filtered out later by the pragmatic module, but most part of the work is done by the syntactic and semantic constraints shown above.

[6] An additional operation of type matching has to be performed after the suffixed verbs are assembled, as in the current version of XMG type copying is performed not via creating a connection between two types (as it is done with attributes), but by copying the value that is there at the moment the operation is performed. To ensure that later type enrichments are copied to all the necessary locations, the class *TypeMatcher* identifies all types of the measure dimensions (M-DIM, NOUN-DIM, VERB-DIM) between the higher and the embedded frames.

4 Discussion

In this paper I have proposed two implementations of analyses that aim to predict the existence and aspect of complex Russian verbs. I have shown that an analysis that is based exclusively on syntactic restrictions (postulating a division of prefixes into several groups) does worse both with respect to precision and recall than an analysis that uses both simple syntactic and semantic restrictions (for the implemented grammar fragment). The summary of precision/recall data is provided in Table 1.

Table 1. Precision, recall and F-measure for two implementations

Analysis	Precision	Recall	F-measure
Frame-based analysis	0,886	1	0.94
Tatevosov (2009)	0,642	0,743	0.689

It is interesting to note that the difference is not huge if one considers only verbs with one or two prefixes and an imperfective suffix added at the last step of the derivation: the number of errors stays close (two versus three within the implemented fragment) and both implementations have full recall with respect to this part of the grammar. The comparison becomes more interesting when we consider the most complex verbs created by the two implementations. The number of models produced here is close: 45 models according to the analysis by Tatevosov (2009) and 49 models in the implementation of the frame semantic analysis. The overlap of these sets constitutes, however, only 27 models. One can argue that such verbs are rare, but I consider them to be an opportunity to test the model, as fitting the theory to such complex cases is not feasible.

Another remark I want to add is that all nine incorrect models that are encountered in the output of the second implementation can be filtered out using basic pragmatic reasoning. Besides, the output of the analysis contains fully spelled-out semantic representations that are obtained compositionally and the interpretation of the prefix in a given position is derived and not stipulated.

In future work I plan to extend the implementation of the frame-based analysis to a larger language fragment and test the predictions of the theory using not only the corpus data and introspection, but setting up experiments to verify the existence of certain complex verbs and then build a database that could be used for future research.

References

Abeillé, A., Rambow, O.: Tree adjoining grammar: an overview. In: Abeillé, A., Rambow, O. (eds.) Tree Adjoining Grammars: Formalisms, Linguistic Analyses and Processing, pp. 1–68. CSLI, Stanford (2000)

Babko-Malaya, O.: Zero morphology: a study of aspect, argument structure, and case. Dissertation, Rutgers University (1999)

Barsalou, L.W.: Frames, concepts, and conceptual fields. In: Lehrer, A., Kittay, E.F. (eds.) Frames, Fields, and Contrasts: New Essays in Semantic and Lexical Organization, Chap. 1, pp. 21–74. Lawrence Erlbaum Associates, Hillsdale (1992)

Crabbé, B., Duchier, D., Gardent, C., Le Roux, J., Parmentier, Y.: XMG: eXtensible MetaGrammar. Comput. Linguist. **39**(3), 591–629 (2013)

Filip, H.: The quantization puzzle. In: Tenny, C.L., Pustejovsky, J. (eds.) Events as Grammatical Objects, pp. 3–60. CSLI Press, Stanford (2000)

Fillmore, C.J.: Frame semantics. In: Linguistics in the Morning Calm, pp. 111–137. Hanshin Publishing Co., Seoul (1982)

Fillmore, C.J., Johnson, C.R., Pertuck, M.R.L.: Background to FrameNet. Int. Lexicogr. **3**(16), 235–250 (2003)

Frank, R.: Syntactic locality and tree adjoining grammar: grammatical, acquisition and processing perspectives. Ph.D. thesis, University of Pennsylvania (1992)

Frank, R.: Phrase Structure Composition and Syntactic Dependencies. MIT Press, Cambridge (2002)

Joshi, A.K., Schabes, Y.: Tree-adjoining grammars. In: Rozenberg, G., Salomaa, A. (eds.) Handbook of Formal Languages, pp. 69–123. Springer, Heidelberg (1997). https://doi.org/10.1007/978-3-642-59126-6_2

Kagan, O.: Scalarity in the Verbal Domain: The Case of Verbal Prefixation in Russian. Cambridge University Press, Cambridge (2015)

Kallmeyer, L., Osswald, R.: A frame-based semantics of the dative alternation in lexicalized tree adjoining grammars. In: Empirical Issues in Syntax and Semantics, p. 9 (2012, to be submitted)

Kallmeyer, L., Osswald, R.: Syntax-driven semantic frame composition in lexicalized tree adjoining grammars. J. Lang. Model. **1**(2), 267–330 (2013)

Lichte, T., Petitjean, S.: Implementing semantic frames as typed feature structures with XMG. J. Lang. Model. **3**(1), 185–228 (2015)

Löbner, S.: Evidence for frames from human language. In: Gamerschlag, T., Gerland, D., Osswald, R., Petersen, W. (eds.) Frames and Concept Types. Studies in Linguistics and Philosophy, vol. 94, pp. 23–67. Springer, Dordrecht (2014). https://doi.org/10.1007/978-3-319-01541-5_2

Petitjean, S., Duchier, D., Parmentier, Y.: XMG 2: describing description languages. In: Logical Aspects of Computational Linguistics (2016)

Ramchand, G.: Time and the event: the semantics of Russian prefixes. Nordlyd **32**(2) (2004)

Romanova, E.: Constructing perfectivity in Russian. Ph.D. thesis, University of Tromsø (2006)

Svenonius, P.: Slavic prefixes and morphology. An introduction to the Nordlyd volume. Nordlyd **32**(2), 177–204 (2004)

Svenonius, P.: Slavic prefixes inside and outside VP. Nordlyd **32**(2), 205–253 (2004)

Tatevosov, S.: Intermediate prefixes in Russian. In: Proceedings of the Annual Workshop on Formal Approaches to Slavic Linguistics, vol. 16 (2007)

Tatevosov, S.: Množestvennaja prefiksacija i anatomija russkogo glagola [Multiple prefixation and the anatomy of Russian verb]. In: Kisseleva, X., Plungian, V., Rakhilina, E., Tatevosov, S. (eds.) Korpusnye issledovanija po russkoj grammatike [Corpus-Based Studies in the Grammar of Russian], pp. 92–156. Probel, Moscow (2009)
Švedova, N.J.: Russkaja Grammatika, vol. 1. Nauka, Moscow (1982)
Zinova, Y.: Russian verbal prefixation. Ph.D. thesis, Heinrich-Heine University, Düsseldorf (2016)
Zinova, Y., Filip, H.: Biaspectual verbs: a marginal category? In: Aher, M., Hole, D., Jeřábek, E., Kupke, C. (eds.) TbiLLC 2013. LNCS, vol. 8984, pp. 310–332. Springer, Heidelberg (2015). https://doi.org/10.1007/978-3-662-46906-4_18

On Generalized Noun Phrases

Richard Zuber(✉)

CNRS, Laboratoire de Linguistique Formelle, Paris, France
Richard.Zuber@linguist.univ-paris-diderot.fr

Abstract. Generalized noun phrases are expressions which play the role of verbal arguments in the same way as ordinary NPs. However proper generalized NPs cannot easily occur in all argumental positions of the verb. Two types of generalized NPs are distinguished and semantically characterized and various properties of functions they denote are studied. These properties indicate similarities and differences between ordinary NPs and generalized NPs and show that generalized NPs essentially extend the expressive power of natural languages.

1 Introduction

Syntactically, generalized noun phrases (GNPs) belong to expressions which typically fulfill the function of argument of the main clause argument like (ordinary) noun phrases (ONPs). However, genuine GNPs are expressions which, in contrast to ONPs, cannot occur in all argumental positions of the verb; in particular they cannot occur in the subject position. Typical examples of (proper) GNPs are anaphoric NPs (ANPs) whose positions in the sentence are determined by the position of their antecedents. The classical example of an ANP belonging to the sub-class of reflexives is the pronoun *himself* and the classical example of the ANP belonging to the sub-class of reciprocals is the pronoun *each other*.

In Sect. 3 we discuss in some details the structure of GNPs and of ANPs in particular. At present it suffices to indicate that we will count as proper GNPs many complex expressions containing *himself* or *each other*. Such complex examples can in particular be Boolean compounds of anaphoric pronouns with anaphoric or non-anaphoric noun phrases. For instance *himself but not most students* is such a reflexive and *each other and ten philosophers* is such a reciprocal.

There are also complex reflexives and reciprocals which are GNPs which are not Boolean compounds. One can obtain such complex ANPs by the application of anaphoric determiners (ADets) to a common noun (CN) (cf. Zuber 2010b). Thus we have reflexive (anaphoric) determiners (RefDets) like for instance *no,... except herself* or *most,..., including Socrates and himself* which can apply to a CN and give complex reflexives like *no teacher, except herself* or *most philosophers, including Socrates and himself.* Similarly reciprocal determiners (RecDets) like *no... except each other, most..., including each other* which can apply to a CN and give complex reflexives and reciprocals like *every logician except each other* (as in *Dan and Leo admire every logician except each other*).

A. Foret et al. (Eds.): FG 2017, LNCS 10686, pp. 142–155, 2018.
https://doi.org/10.1007/978-3-662-56343-4_9

Another important class of GNPs, distinct from ANPs representing nominal anaphors, is formed by comparative generalised NPs, CNPs. It contains two sub-classes. First, to CNPs belong what Keenan (2016) calls *predicate anaphors*, that is expressions like *more linguists than Dan* or *the greatest number of teachers* (as found in *Leo met more linguists than Dan/the geatest number of teachers*). Clearly such expressions can be used, at least "at the surface" as verbal arguments for some verbal positions.

The second sub-class of CNPs is formed from what we will call, for reasons to be explained below, higher order comparative GNPs (HCNPs). These are expressions like *the same books* or *very different articles and the oldest book in the library.*

We will also discuss HCNPs formed with determiner like *the same number of.* They give rise to the following HCNPs: *the same number of students, almost the same number of articles*, etc.

The following examples show that the indicated reflexives, reciprocals and HCNPs are indeed genuine GNPs:

(1) *He(self) admires Dan.

(2) *Each other admires Dan and Leo.

(3) *Most philosophers, including Socrates and himself admire Dan.

(4) *Every logician except each other admire Dan and Leo.

(5) *More students than Dan knows Leo.

(6) *The same articles and the oldest book in the library read Dan and Leo.

Even though many HCNPs can occur in the subject position (for instance *the same CN* as in *the same actors played three characters in the movie*) we will consider them, for formal reason to be given below, as genuine GNPs.

The purpose of this paper is to characterize in a preliminary way denotations of GNPs in their opposition to ONPs. We will be mainly interested in formal properties of functions denoted by GNPs. In the next section we recall some basic notions from the generalized quantifier theory and, more importantly, we show how they can be extended so that they apply to denotations of GNPs, more specifically to CNPs, ANPs and HCNPs. In Sect. 3 we indicate various syntactic forms which GNPs can take. In Sect. 4 we give semantics of some basic GNPs and indicate properties of functions representing this semantics.

2 Formal Preliminaries

We will consider binary relations and functions over a universe E, assumed to be finite throughout this paper. $D(R)$ denotes the domain of R. The relation I is the identity relation: $I = \{\langle x, y\rangle : x = y\}$. If R is a binary relation and X a set, then $R/X = R \cap (X \times X)$. The binary relation R^S is the greatest symmetric relation included in R, that is $R^S = R \cap R^{-1}$ and $R^{S-} = R^S \cap I'$. If R is an irreflexive symmetric relation (i.e. $R \cap R^{-1} \cap I = \emptyset$) then $\Pi(R)$ is the least fine partition of R such that every one of its blocks is of the form $(A \times A) \cap I'$. A partition is *trivial* iff it contains only one block. Observe that if R is an irreflexive symmetric relation and $\Pi(R)$ is not trivial than every block of $\Pi(R)$ contains at least two elements.

If a function takes only a binary relation as argument, its type is noted $\langle 2 : \tau \rangle$, where τ is the type of the output; if a function takes a set and a binary relation as arguments, its type is noted $\langle 1, 2 : \tau \rangle$. If $\tau = 1$ then the output of the function is a set of individuals and thus its type is $\langle 2 : 1 \rangle$ or $\langle 1, 2 : 1 \rangle$. The function $SELF$, denoted by the reflexive *himself* and defined as $SELF(R) = \{x : \langle x, x \rangle \in R\}$, is of type $\langle 2 : 1 \rangle$ and the function denoted by the anaphoric determiner *every...but himself* is of type $\langle 1, 2 : 1 \rangle$. We will consider here also the case when τ corresponds to a set of type $\langle 1 \rangle$ quantifiers and thus τ equals, in Montagovian notation, $\langle \langle \langle e, t \rangle t \rangle t \rangle$. The type of such functions will be noted either $\langle 2 : \langle 1 \rangle \rangle$ - functions from binary relations to sets of type $\langle 1 \rangle$ quantifiers or $\langle 1, 2 : \langle 1 \rangle \rangle$ - functions from sets and binary relations to sets of type $\langle 1 \rangle$ quantifiers.

Basic type $\langle 1 \rangle$ quantifiers are functions from sets to truth-values. Functions from sets to type $\langle 1 \rangle$ quantifiers are type $\langle 1, 1 \rangle$ quantifiers which are denoted by (nominal) unary determiners. Basic type $\langle 1 \rangle$ quantifiers are denotations of subject NPs. However, NPs can also occur in the direct object position and in this case their denotations do not take sets (denotations of VPs) as arguments but denotations of TVPs (relations) as arguments (Keenan 2016). To account for this eventuality the domain of application of basic type $\langle 1 \rangle$ quantifiers is extended in the way that it contains in addition the set of binary relations. When a quantifier Q acts as a "direct object" we get its *accusative case extension* Q_{acc} (Keenan and Westerståhl 1997):

Definition 1. *For each type* $\langle 1 \rangle$ *quantifier* Q, $Q_{acc}R = \{a : Q(aR) = 1\}$, *where* $aR = \{y : \langle a, y \rangle \in R\}$.

Formally accusative extensions of type $\langle 1 \rangle$ quantifiers are of the same expressive power as type $\langle 1 \rangle$ quantifiers because the algebra of type $\langle 1 \rangle$ quantifiers is isomorphic to the algebra of the accusative extensions of type $\langle 1 \rangle$ quantifiers.

Various applications of the notion of the accusative extension of a quantifier are given in Keenan (2016) where in particular it is shown that the accusative extension allows us to avoid recourse to **LF** movement when interpreting NPs in the object position.

A type $\langle 1 \rangle$ quantifier Q is *positive*, $Q \in POS$, iff $\emptyset \notin Q$; Q is *natural* iff either $Q \in POS$ and $E \in Q$ or $Q \notin POS$ and $E \notin Q$; Q is plural, $Q \in PL$, iff if $X \in Q$ then $|X| \geq 2$. Q_A is the atomic quantifier true of just A.

A special class of type $\langle 1 \rangle$ quantifiers is formed by *individuals*: I_a is an individual (generated by $a \in E$) iff $I_a = \{X : a \in X\}$. They are denotations of proper names. More generally, $Ft(A)$, the *(principal) filter generated by the set* A, is defined as $Ft(A) = \{X : X \subseteq E \wedge A \subseteq X\}$. NPs of the form *Every CN* denote principal filters generated by the denotation of *CN*. Meets of two principal filters are principal filters: $Ft(A) \cap Ft(B) = Ft(A \cup B)$.

We will use also the property of *living on* (cf. Barwise and Cooper 1981). The basic type $\langle 1 \rangle$ quantifier lives on a set A (where $A \subseteq E$) iff for all $X \subseteq E$, $Q(X) = Q(X \cap A)$. We extend the notion of living on to the type $\langle 2:1 \rangle$ functions. Thus a type $\langle 2:1 \rangle$ function F lives on the relation S iff $F(R) = F(R \cap S)$ for any binary relation R. It is easy to see that Q lives on A iff Q_{acc} lives on $E \times A$.

If E is finite then there is always a smallest set on which a quantifier Q lives. If A is a set on which Q lives we will write $Li(Q, A)$ and the smallest set on which Q lives will be noted $SLi(Q)$. A related notion is the notion of a witness set of the quantifier Q, relative to the set A on which Q lives:

Definition 2. $W \in Wt_Q(A)$ *iff* $W \in Q \wedge W \subseteq A \wedge Li(Q, A)$.

One can see that any principal filter lives on the set by which it is generated, and, moreover, this set is its witness set. Atomic quantifiers live on the universe E only.

Accusative extensions of type $\langle 1 \rangle$ quantifiers are specific type $\langle 2:1 \rangle$ functions. They satisfy the invariance condition called *accusative extension condition* **EC** (Keenan and Westerståhl 1997):

Definition 3. *A type* $\langle 2:1 \rangle$ *function* F *satisfies* ***EC*** *iff for* R *and* S *binary relations, and* $a, b \in E$, *if* $aR = bS$ *then* $a \in F(R)$ *iff* $b \in F(S)$.

Observe that if F satisfies **EC** then for all $X \subseteq E$ either $F(E \times X) = \emptyset$ or $F(E \times X) = E$. Given that $SELF(E \times A) = A$ the function $SELF$ does not satisfy **EC**. The function $SELF$ satisfies the following weaker predicate invariance condition **PI** (Keenan 2007):

Definition 4. *A type* $\langle 2:1 \rangle$ *function* F *is predicate invariant (**PI**) iff for* R *and* S *binary relations, and* $a \in E$, *if* $aR = aS$ *then* $a \in F(R)$ *iff* $a \in F(S)$.

This condition is also satisfied for instance by the function $ONLY\text{-}SELF$ defined as follows: $ONLY\text{-}SELF(R) = \{x : xR = \{x\}\}$. Given that $ONLY\text{-}SELF(E \times \{a\}) = \{a\}$, the function $ONLY\text{-}SELF$ does not satisfy **EC**.

The following proposition indicates another way to define **PI** (Zuber 2016):

Proposition 1. *A type* $\langle 2:1 \rangle$ *function* F *is predicate invariant iff for any* $x \in E$ *and any binary relation* R, $x \in F(R)$ *iff* $x \in F(\{x\} \times xR)$.

The **PI** condition is weaker than **EC**. The function $MORE_{S,d}$ which interprets the CNP *more students than Dan* and which is defined as $MORE_{S,d}(R) = \{x : |xR| > |dR|\}$ satisfies another weakening of **EC**, the so-called *argument invariance* condition **AI** (Keenan and Westerståhl 1997):

Definition 5. *A type* $\langle 2:1 \rangle$ *function* F *is argument invariant (**AI**) iff for any binary relation* R *and* $a, b \in E$, *if* $aR = bR$ *then* $a \in F(R)$ *iff* $b \in F(R)$.

The invariant conditions **EC**, **PI** and **AI** concern type $\langle 2:1 \rangle$ functions, considered here as being denoted by full GNPs. As an illustration we provide a similar definition for type $\langle 1, 2:1 \rangle$ functions denoted by ordinary (nominal) determiners. Thus:

Definition 6. *A type* $\langle 1,2 : 1\rangle$ *function* F *satisfies* ***D1EC*** *iff for* R *and* S *binary relations,* $X \subseteq E$ *and* $a, b \in E$*, if* $aR \cap X = bS \cap X$ *then* $a \in F(X,R)$ *iff* $b \in F(X,S)$.

Observe that if $F(X,R)$ satisfies **D1EC** then for all $X, A \subseteq E$ either $F(X, E\times A) = \emptyset$ or $F(X, E\times A) = E$. Denotations of ordinary determiners occurring in NPs which take direct object position satisfy **D1EC**. More precisely, if D is a type $\langle 1,1\rangle$ conservative quantifier, then the function $F(X,R) = D(X)_{acc}(R)$ satisfies **D1EC**: in this case $F(X,R) = \{y : D(X)(yR \cap X) = 1\}$ and $F(X,S) = \{y : D(X)(yS \cap X) = 1\}$. So if $aR \cap X = bS \cap X$ then $a \in F(X,R)$ iff $b \in F(X,S)$.

The above invariance principles concern type $\langle 2 : 1\rangle$ and type $\langle 1,2 : 1\rangle$ functions. We need to present similar "higher order" invariance principles for type $\langle 2 : \langle 1\rangle\rangle$ functions, that is functions having as output a set of type $\langle 1\rangle$ quantifiers.

One can distinguish various kinds of type $\langle 2 : \langle 1\rangle\rangle$ functions. Observe first that any type $\langle 2 : 1\rangle$ function whose output is denoted by a VP can be lifted to a type $\langle 2 : \langle 1\rangle\rangle$ function. The accusative extension of a type $\langle 1\rangle$ quantifier Q can be lifted to type $\langle 2 : \langle 1\rangle\rangle$ function in the way indicated in (7). Such functions will be called *accusative lifts*. More generally, if F is a type $\langle 2 : 1\rangle$ function, its lift F^L, a type $\langle 2 : \langle 1\rangle\rangle$ function, is defined in (8):

(7) $Q^L_{acc}(R) = \{Z : Z(Q_{acc}(R)) = 1\}$.

(8) $F^L(R) = \{Z : Z(F(R)) = 1\}$.

The variable Z above runs over the set of type $\langle 1\rangle$ quantifiers.

For type $\langle 2 : \langle 1\rangle\rangle$ functions which are lifts of type $\langle 2 : 1\rangle$ functions we have:

Proposition 2. *If a type* $\langle 2 : \langle 1\rangle\rangle$ *function* F *is a lift of a type* $\langle 2 : 1\rangle$ *function then for any type* $\langle 1\rangle$ *quantifiers* Q_1 *and* Q_2 *and any binary relation* R*, if* $Q_1 \in F(R)$ *and* $Q_2 \in F(R)$ *then* $(Q_1 \wedge Q_2) \in F(R)$.

For type $\langle 2 : \langle 1\rangle\rangle$ functions which are accusative lifts we have:

Proposition 3. *Let* F *be a type* $\langle 2 : \langle 1\rangle\rangle$ *function which is an accusative lift. Then for any* $A, B \subseteq E$*, any binary relation* R*,* $Ft(A) \in F(R)$ *and* $Ft(B) \in F(R)$ *iff* $Ft(A \cup B) \in F(R)$.

Accusative lifts satisfy the following higher order extension condition **HEC** (Zuber 2014):

Definition 7. *A type* $\langle 2 : \langle 1\rangle\rangle$ *function* F *satisfies* ***HEC*** *(higher order extension condition) iff for any natural type* $\langle 1\rangle$ *quantifiers* Q_1 *and* Q_2 *with the same polarity, any* $A, B \subseteq E$*, any binary relations* R, S*, if* $Li(Q_1, A)$*,* $Li(Q_2, B)$ *and* $\forall_{a\in A}\forall_{b\in B}(aR = bS)$ *then* $Q_1 \in F(R)$ *iff* $Q_2 \in F(S)$.

Functions satisfying **HEC** have the following property:

Proposition 4. *Let* F *satisfies* ***HEC*** *and let* $R = E \times C$*, for* $C \subseteq E$ *arbitrary. Then for any* $X \subseteq E$ *either* $Ft(X) \in F(R)$ *or for any* X*,* $Ft(X) \notin F(R)$.

Thus a function satisfying **HEC** condition and whose argument is the cross-product relation of the form $E \times A$, has in its output either all principal filters or no principal filter. We will see that the function denoted by the ANP *each other* does not satisfy **HEC**.

It follows from Proposition 4 that lifts of genuine predicate invariant functions do not satisfy **HEC**. They satisfy the following weaker condition (Zuber 2014):

Definition 8. *A type* $\langle 2 : \langle 1 \rangle \rangle$ *function* F *satisfies* ***HPI*** *(higher order predicate invariance) iff for type* $\langle 1 \rangle$ *quantifier* Q*, any* $A \subseteq E$*, any binary relations* R, S*, if* $Li(Q, A)$ *and* $\forall_{a \in A}(aR = aS)$ *then* $Q \in F(R)$ *iff* $Q \in F(S)$.

The higher order property corresponding to **AI** is the higher order argument invariance:

Definition 9. *A type* $\langle 2 : \langle 1 \rangle \rangle$ *function* F *satisfies* ***HAI*** *(higher order argument invariance) iff for any natural type* $\langle 1 \rangle$ *quantifiers* Q_1 *and* Q_2 *with the same polarity, any* $A, B \subseteq E$*, any binary relation* R*, if* $SLi(Q_1, A)$*,* $SLi(Q_2, B)$ *and* $\forall_{a \in A} \forall_{b \in B}(aR = aS)$ *then* $Q_1 \in F(R)$ *iff* $Q_2 \in F(R)$.

Higher order invariance conditions are generalizations of "simple" invariance conditions because it can be shown (cf. Zuber 2014) that lifts of functions satisfying simple invariance condition satisfy higher order invariance conditions. Thus the accusative lift of a type $\langle 1 \rangle$ quantifier satisfies **HEC**, the lift a function satisfying **PI** satisfies **HPI** and the lift of a function satisfying **AI** satisfies **HAI**.

3 Structure of Generalized Noun Phrases

In this section we indicate some structural and syntactic differences and similarities between ONPs and GNPs by comparing their respective structures. The remarks which follow are not intended, however, to characterise syntactically the class of proper GNPs. Moreover, we have to keep in mind that we consider that the class of GNPs is strictly included in the class of NPs and thus that there are genuine, or proper, GNPs which are not ONPs. In the next section we will characterise semantically two classes of proper GNPs: simple and higher order GNPs. Roughly speaking, simple GNPs are GNPs related to reflexives or simple comparatives (or predicate anaphors) and higher order GNPs are those which are related to reciprocals or HCGNPs such as *the same CN*.

The first thing to notice is that among genuine GNPs there are no elements corresponding to proper names, which, obviously are OMPs. Thus there are no morphologically simple non-pronominal genuine GNPs. We observe that morphologically simple or "almost simple" genuine GNPs have a pronominal character. This is the case with the reflexive *himself* or reciprocal *each other*. Interestingly "ordinary" pronouns are ONPs.

One of very productive ways of forming complex ONPs is by the application of determiners to CNs. Thus there is a natural class of ONPs which are of the form *Det CN* where *Det* is an unary determiner that is a functional expression which when applied to one CN gives an NP. Such "ordinary" determiners have been extensively studied, various sub-classes of them have been distinguished and formal properties of their denotations, that is type $\langle 1, 1\rangle$ quantifiers, have been established. It is generally admitted that unary determiners denote conservative type $\langle 1, 1\rangle$ functions and conservative functions have formally important sub-classes of intersective and co-intersective functions. For instance the determiner *most* denotes a conservative function which is conservative but neither intersective nor co-intersective, the numerals are determiners which denote intersective functions and determiners like *every* or *every...but ten* denote co-intersective functions.

Now, important point is that there is also a class of genuine GNPs which are obtained by the application of a (generalised) determiner to a CN, that is GNPs of the form *GDet CN*, where *GDet*, a generalized determiner, is a functional expression which when applied to a CN gives a genuine GNP. GDets in their turn can be divided into anaphoric GDets (ADets) and comparative GDets (CGDets). Finally, among ADets we have RefDets, reflexive determiners and RecDets, that is reciprocal determiners. To see these different classes of GDets consider the following examples:

(9) Dan hates every linguist except himself

(10) Dan knows more linguists than Leo

(11) Leo and Dan admire no linguist except each other

(12) Leo and Dan read the same books

In (9) the determiner *every... except himself* is a RefDet. Similarly *no...except himself and Dan* and *most..., including himself* are RefDets. In (10) we have a CDet *more... than Leo*. In (11) the expression *no... except each other* is a RecDet as are for instance expressions like *every... except each other and Dan* or *most, including each other*. In (12) we have a CDet *the same*. Other examples of such determiners are represented by *different, very different, quite different, similar, very similar, almost the same*, etc.

Another very productive way of forming NP is by the use of Boolean connectors. For instance the following NPs are such Boolean compounds *Dan and most students, ten logicians and some linguists, five students and no teacher except Dan*. As the following examples show there are also genuine GNPs which are Boolean compounds:

(13) Dan admires himself and most philosophers

(14) Leo and Dan admire each other but not themselves

(15) Leo and Dan read five articles and the same books

(16) Leo and Dan read the same articles and different books

(17) Leo and Dan admire each other but not themselves and Lea

(18) Dan and Leo admire each other, themselves and the same linguists

Observe that in the above examples GNPs belonging to various subclasses are conjoined. This does not mean that all GNPs can be freely conjoined in the same way as ONPs cannot always be freely conjoined.

The next similarity between ONPs and genuine GNPs I mention briefly concerns the possibility of their modification by the so-called categorically polyvalent modifiers, CPM, that is modifiers which can apply to expressions of different categories. CPM include expressions like *only, also, even, at least*, etc. These modifiers can modify expressions of various categories and in particular they can apply to ONPs since we have: *even Dan, only Dan and Bill, also some students, at least ten teachers*, etc. As the following examples show, proper GNPs can also be modified by the CPM:

(19) Leo admires only/even/at/least himself.

(20) Leo and Dan admire at least/at most/only/even each other.

(21) They read even/at least the same books.

Finally, ONPs and proper GNPs can also play the role of arguments of non-verbal predicatives, that is complex expressions which are not modifiers and do not contain a verb but which take GNPs as an argument. A natural class of such predicatives is formed either from transitive CNs (like *friend of* or *young grand-parent of*) or from transitive adjectives (like *jealous of*). Thus we have the following predicatives in which ONPs occur as arguments: *grand-parents of ten children, friends of some gangsters*. Proper GNPs can also occur in such contexts since we have: *grand-parents of the same students, fond of the same students/themselves* or *jealous of each other*.

Proper GNPs can also occur in relative clauses and other embedding constructions. In this case, however, the complex constructions containing such embedded GNPs can easily occur in the subject position: the expressions *persons with the same taste/who admire each other* and *to hate each other* can be used as subject NPs. In addition both types of noun phrases can occur as arguments in prepositional phrases since we have *leave with each other* or *talk with the same persons*.

A general form of sentences in which ANPs, CNPs and HCNPs occur and which we will consider here, is given in (22):

(22) *NP TVP GNP*

TVP is a transitive verb phrase which denotes a binary relation and GNP is either *himself* or *each other* or has the form RefDet(CN), RecDet(CN), *the same CN* or is a Boolean combination of all such cases.

4 Formal Properties of Generalized Noun Phrases

In this section we analyze properties of full GNPs and not of their specific parts such as reflexive determiners, comparative determiners or reciprocal determiners.

Properties of RefDets are studied in Zuber (2010b) and comparative determiners - in Zuber (2011). A proposal to treat some higher order comparatives is given in Zuber (2017b).

The fact that nominal anaphors have various properties which distinguish them from ONPs is well known. For instance Geach (1968) indicates various peculiarities of the pronoun *himself* and suggest that some of them have been discussed by medieval philosophers. I will illustrate first informally some differences between ONPs and various simple proper GNPs, using in particular various observations from Keenan (2007, 2016) and Zuber (2011).

When an ONP (in the subject position) denotes a type $\langle 1 \rangle$ quantifier Q then when it occurs in the object position it denotes the accusative extension Q_{acc} of Q. Accusative extensions of a quantifier satisfy **EC**. This means that if, for instance, the persons that Leo washes are the same as the persons that Dan shaves than the following two sentence forms have the same truth values, for any NP:

(23) Leo washes NP

(24) Dan shaves NP

This is not the case with functions denoted by simple proper GNPs. Consider for instance the reflexive *himself*. Suppose again that persons that Dan washes are the same as the persons that Leo shaves. In this case the following sentences can fail to have the same truth value:

(25) Dan washes himself.

(26) Leo shaves himself.

Consider now the CNP *the greatest number of languages*. Suppose that the set of languages that Dan speaks is the same as the set of languages that Leo studies. It does not follow from this that the following sentences have the same truth value:

(27) Dan speaks the greatest number of languages.

(28) Leo studies the greatest number of languages.

Thus simple GNPs denote functions which do not satisfy **EC** satisfied by accusative extensions denoted by ONPs in the object position. Functions denoted by reflexives satisfy the weaker **PI** condition and functions denoted by simple comparatives satisfy the weaker **AI** condition.

Higher order GNPs are additionally different from ONPs and from simple GNPs. To see this informally consider the following examples (cf. Zuber 2014):

(29) a. Leo and Lea hug each other/read the same books.

b. Bill and Sue hug each other/read the same books.

(30) Leo, Lea, Bill and Sue hug each other/read the same books.

Clearly (29a) in conjunction with (29b) does not entail (30). However, if we replace *each other* or *the same books* by an ordinary NP or by a simple proper GNP, the corresponding entailment holds. This means that the conjunction *and* cannot be understood pointwise and that the functions denoted by GNPs like *the same CN* and *each other* are of not the same type as functions denoted by ONPs or by simple GNPs.

Observe that the non entailment of (30) from (29a) and (29b) in conjunction with Proposition 3 indicates that the GNPs *the same books* and *each other* are not accusative lifts (of any type $\langle 1 \rangle$ quantifier).

The question one can ask now is what is the logical type of the result of the function denoted by GNPs which are reciprocals or HCNPs. We know that sentences with such GNPs (in the object position) do not take proper nouns as subjects and thus the type of objects denoted by the subject NP cannot be e, the type corresponding to individuals. We can suppose that it is of the raised type $\langle\langle e,t\rangle,t\rangle$, which, ignoring directionality, corresponds to the category *S/(S/NP)*. Since *the same CN* and *each other* form a verbal argument playing the role of direct object, *the same (CN)* and *each other* apply to a transitive verb to form a VP. Semantically, this verb phrase denotes a set of type $\langle 1 \rangle$ quantifiers. Thus, in order to avoid the type mismatch, the verb phrase must be raised to become of the category *S/(S/(S/NP))*. This can be done using the following higher order reduction *via* function application (where "+" symbolises the function application):

(31) $S/(S/NP) + S(S/(S/NP)) = S$

Thus in (31) the VP has been raised to the category *S(S/(S/NP))* whose type is now $\langle\langle\langle e,t\rangle t\rangle t\rangle$. This means that such raised VPs denote a set of type $\langle 1 \rangle$ quantifiers and consequently the sentence of the form (22) is true iff the quantifier denoted by the NP belongs to the set denoted by the $TVP\ GNP$. Keenan and Faltz (1985) show that (extensional) non-raised VPs (classically) denote, up to the isomorphism, specific characteristic functions of sets of type $\langle 1 \rangle$ quantifiers, that is they denote a set of quantifiers. The reason is that these characteristic functions are in addition homomorphisms (from the algebra of quantifiers to the algebra of truth-values) and the algebra of such homomorphic functions is isomorphic to the algebra of sets (subsets of the universe), the classical denotational domain of VPs (or of one-place predicates in the first order logic). Thus "classical" denotations of VPs are homomorphic in the sense that they preserve meets in particular. We have seen that this is not the case for the VPs formed from higher order GNPs given the non-entailment between (29) and (30) and consequently VPs with higher order GNPs do not denote sets, subsets of E, but sets of type $\langle 1 \rangle$ quantifiers.

Let me start the discussion of formal properties by the following:

Proposition 5. *Boolean algebra of type $\langle 1 \rangle$ quantifiers is isomorphic to the algebra of intersective type $\langle 1,1 \rangle$ quantifiers and to the algebra of co-intersective quantifiers.*

The proof of this proposition is obvious if one observes that there is one to one correspondence between atoms of the algebra of type $\langle 1\rangle$ quantifiers and the atoms of the algebra of intersective quantifiers. Indeed for any set A the singleton $\{A\}$ is an atomic type $\langle 1\rangle$ quantifier and the type $\langle 1,1\rangle$ quantifier F_A such that $F_A(X)(Y) = 1$ iff $X \cap Y = A$ is an atom of the algebra of intersective quantifiers.

Thus there are as many type $\langle 1\rangle$ quantifiers as there are intersective quantifiers. Since (cf. Keenan and Westerståhl 1997) any type $\langle 1\rangle$ quantifier is expressible (in English) by an ONP (of English) this means that there as many (English) ONPs as there are intersective (or co-intersective) quantifiers. If the universe E has n elements then there are 2^k, for $k = 2^n$ type $\langle 1\rangle$ quantifiers. But there are much more anaphoric type $\langle 2:1\rangle$ functions. As Keenan (2007) indicates in this case there are 2^m, for $m = n \times 2^n$ functions satisfying **PI**. The number off all functions from the set of binary relations to the set of sets equals k^l for $k = 2^n$ and $l = 2^{n\times n}$. This means that in the universe with just two elements there are 16 type $\langle 1\rangle$ quantifiers, 256 functions satisfying **PI** and 4^{16} functions from binary relations to sets.

Let us now define some functions denoted by some GNPs. To define the type $\langle 2:\langle 1\rangle\rangle$ function EA denoted by the reciprocal *each other* we use the partition $\Pi(R^{S-})$ (Zuber 2016). Our definition is the definition "by cases" which depend on whether the partition $\Pi(R^{S-})$ is trivial or non-trivial. Thus

Definition 10.

(i) $EA(R) = \{Q : Q \in PL \wedge \neg 2(E) \subseteq Q\}$ *if* $R^{S-} = \emptyset$
(ii) $EA(R) = \{Q : Q \in PL \wedge Q_{D(B)} \subseteq Q\}$, *if* $\Pi(R^{S-})$ *is trivial with* B *as its only block*
(iii) $EA(R) = \{Q : Q \in PL \wedge \exists_B(B \in \Pi(R^{S-}) \wedge Q(D(B) = 1\} \cup \{Q : Q \in PL \wedge \exists_B(B \in \Pi(R^{S-}) \wedge Q = \neg Q_{D(B)}\}$ *if* $\Pi(R^{S-})$ *is non-trivial.*

Functions denoted by GNPs are anaphoric in the sense that they satisfy predicate invariance conditions **PI** or **HPI** and do not satisfy stronger conditions **EC** or **HEC**. We have already seen that $SELF$ and $ONLY\text{-}SELF$ are anaphoric in that sense. Using Proposition 5 and Definition 8 we show that the function EA in definition (10) is anaphoric (because for $R = E \times A$ the partition $\Pi(R^{S-})$ is trivial). Some other higher order anaphoric functions are discussed in Zuber (2016, 2017a). In particular RefDets and RecDets and properties of functions they denote are discussed in Zuber (2017a).

To define the functions $SAME(X,R)$ and $SAME\text{-}N$ denoted by *the same CN* and *the same number of CN* respectively, where *CN* denotes X, we will use the set partitions defined by the following equivalence relations (Zuber 2017b):

Definition 11.

(i) $e_R = \{\langle x,y\rangle : xR = yR\}$
(ii) $e_{R,n} = \{\langle x,y\rangle : card(xR) = card(yR)\}$.

We will say that the block of a partition is singular if it is a singleton. A block B is plural, $B \in PL$, if it is contains at least two elements. A partition is atomic iff all its blocks are singular. With the help of these notions, using the partition

$\Pi_{R_A}(E)$ we can now express the function $SAME(X,R)$, where R is a binary relation, as follows (for X and R non-empty and where R_X is a subrelation of R whose range is restricted to X):

Definition 12. $SAME(X,R) =$
(i) $= \{Q : Q \in PLR \wedge \neg 2(E) \subseteq Q\}$, *if* $\Pi_{R_X}(E)$ *is atomic*
(ii) $= \{Q : Q \in PLR \wedge \exists_B (B \in PL \wedge B \in \Pi_{R_X}(E) \wedge Q(B) = 1\}$
$\cup \{Q : Q \in PLR \wedge \exists_{C \subseteq E} \forall_{B \in \Pi_{R_X}(E)} C \not\subseteq B) \wedge \neg ALL(C) \subseteq Q)\}$, *if* $\Pi_{R_X}(E)$ *is not atomic.*

The above definition says that $SAME$ applied to a set X and a binary relation R gives as result a set of quantifiers. This set can be decomposed into various subsets depending on the structure of the partition of E induced by R and X. Clause (i) says that when the partition is atomic then no two objects are in the relation R with all objects of a sub-set of X. This entails that the quantifier denoted by *no two objects* and any of its consequences belong to the set $SAME(X,R)$. This means that, for instance, the quantifiers denoted by *no five objects* or *no two students* also belong to the set $SAME(X,R)$.

Clause (ii) concerns the case where the partition is not atomic. In this case there is at least one plural block of the partition such that all its members are, roughly speaking, in the relation R with the same subset of X. This block corresponds to the property expressing the sameness we are looking for and a plural quantifier can be true or false of it. The second part of the clause (ii) provides a set of quantifiers obtained from a "negative information" given by sets which are not blocks of the partition. If, for instance, Jiro and Taro are Japanese students who read different books then no set to which they belong is a block of $\Pi_{R_B}(E)$ - where R corresponds to $READ$ and B - to $BOOK$. Then, according to the second part of the clause (ii), the quantifiers denoted by the *NPs not all Japanese students*, *not all students* and *not all Japanese* belong to $SAME(B,R)$.

The definition of the function $SAME$-N denoted by the generalized determiner *the same number of* is quite similar to the definition of the function $SAME(X,R)$. We just have to replace everywhere in Definition 12 the partition $\Pi_{R_X}(E)$ by the partition $\Pi_{R_{X,n}}(E)$. Consequently we have:

Definition 13. $SAME\text{-}N(X,R) =$
(i) $= \{Q : Q \in PLR \wedge \neg 2(E) \subseteq Q\}$, *if* $\Pi_{R_{X,n}}(E)$ *is atomic*
(ii) $= \{Q : Q \in PLR \wedge \exists_B (B \in PL \wedge B \in \Pi_{R_{X,n}}(E) \wedge Q(B) = 1\}$
$\cup \{Q : Q \in PLR \wedge \exists_{C \subseteq E} (C \notin \Pi_{R_X}(E) \wedge \neg ALL(C) \subseteq Q)\}$, *if* $\Pi_{R_X}(E)$ *is not atomic.*

Definitions 12 and 13 provide the readings of *the same* and *the same number of* without the existential import that is without the presupposition that the set denoted by *CN* is not empty. In order to get the reading in which the existential import is involved the following equivalence relations have to be used:

Definition 14. $e_R^{ei} = \{\langle x,y \rangle : (xR = yR \wedge xR \neq \emptyset) \vee (x = y)\}$.

Definition 15. $e^{ei}_{R,n} = \{\langle x, y\rangle : (|xR| = |yR| \wedge xR \neq \emptyset) \vee (x = y)\}$.

The relation e^{ei}_R defines the partition $\Pi^{ei}_R(E)$ and the relation $e^{ei}_{R,n}$ defines the partition $\Pi^{ei}_{R,n}(E)$. It follows from Definitions 14 and 15 that if $aR = \emptyset$, then the singleton $\{a\}$ is a singular block of both partitions Π^{ei}_R and $\Pi^{ei}_{R,n}$ and thus is not a member of any plural quantifier. Consequently the reading of *the same* with the existential import is given in Definition 16:

Definition 16. $SAME^{ei}(X, R) =$
(i) $= \{Q : Q \in PLR \wedge \neg 2(E) \subseteq Q\}$, *if* $\Pi^{ei}_{R_X}(E)$ *is atomic*
(ii) $= \{Q : Q \in PLR \wedge \exists_B(B \in PL \wedge B \in \Pi^{ei}_{R_X}(E) \wedge Q(B) = 1\}$
$\cup \{Q : Q \in PLR \wedge \exists_{C\subseteq E}\forall_{B\in\Pi^{ei}_{R_X}(E)} C \not\subseteq B) \wedge \neg ALL(C) \subseteq Q)\}$, *if* $\Pi^{ei}_{R_X}(E)$ *is not atomic.*

It is easy, though tedious, to show that functions defined in Definitions 12, 13 and 16 satisfy **HAI** (and do not satisfy **HEC**). Consequently, higher order GNPs also denote functions which are not accusative extensions of type $\langle 1\rangle$ quantifiers.

5 Conclusive Remarks

Generalized noun phrases are expressions which, syntactically play the role of direct objects as do ordinary NPs. Semantically, however, they do not denote type $\langle 1\rangle$ quantifiers or their accusative extensions. Functions they denote satisfy weaker conditions than the extension condition, which is satisfied by accusative extensions of type $\langle 1\rangle$ quantifiers. In spite of that they resemble quantifiers in various ways.

We distinguished two types of GNPs, according to the type of functions they denote: simple GNPs (for instance reflexives and predicate anaphors) denote type $\langle 2:1\rangle$ functions and higher order GNPs (like reciprocals) denote type $\langle 2:\langle 1\rangle\rangle$ functions. Both types of these functions satisfy similar invariance conditions and both types of GNPs have their syntactic structure similar to the structure of ONPs.

Syntactic similarity in the structures of GNPs and ONPs and the fact that the two types of expressions, ONPs and GNPs can occur as different conjuncts in the same Boolean compounds indicates that GNPs should not be considered as a new syntactic category. Rather, to account for the specificity of their semantics we should consider, in the spirit of Partee et al. (1986) that the type of NPs can change depending on the environment it finds itself in. In this case higher order GNPs give rise to the VP raising.

Formal properties of GNPs presented in this paper shows that the existence of anaphors and higher order comparative NPs strongly extends the expressive power of NLs. Keenan (2007, 2016) shows that denotations of reflexive anaphors and predicate anaphors lie outside the class of classically defined generalized quantifiers (they do not satisfy the extension condition). Results presented in this paper show that in addition higher order GNPs form non-homomorphic predicates which force the VP raising because their denotations are not lifts of type $\langle 2:1\rangle$ functions.

References

Barwise, J., Cooper, R.: Generalized quantifiers and natural language. Linguist. Philos. **4**, 159–219 (1981)

Geach, P.T.: Reference and Generality. Cornell University Press, Ithaca (1968)

Keenan, E.L.: On the denotations of anaphors. Res. Lang. Comput. **5**, 5–17 (2007)

Keenan, E.L.: In situ interpretation without type mismatches. J. Seman. **33**(1), 87–106 (2016)

Keenan, E., Faltz, L.: Boolean Semantics for Natural Language. Reidel Publishing Company, Dordrecht (1985)

Keenan, E.L., Westerståhl, D.: Generalized quantifiers in linguistics and logic. In: van Benthem, J., ter Meulen, A. (eds.) Handbook of Logic and Language, pp. 837–893. Elsevier, Amsterdam (1997)

Partee, B.: Noun phrase interpretation and type-shifting principles. In: Groenendijk, J., et al. (eds.) Studies in Discourse Representation Theory and Theory of Generalised Quantifiers, pp. 115–143. Forris, Dordrecht (1986)

Zuber, R.: Semantic constraints on anaphoric determiners. Res. Lang. Comput. **8**, 255–271 (2010b)

Zuber, R.: Some generalised comparative determiners. In: Pogodalla, S., Prost, J.-P. (eds.) LACL 2011. LNCS (LNAI), vol. 6736, pp. 267–281. Springer, Heidelberg (2011). https://doi.org/10.1007/978-3-642-22221-4_18

Zuber, R.: Generalising predicate and argument invariance. In: Asher, N., Soloviev, S. (eds.) LACL 2014. LNCS, vol. 8535, pp. 163–176. Springer, Heidelberg (2014)

Zuber, R.: Anaphors and quantifiers. In: Väänänen, J., Hirvonen, Å., de Queiroz, R. (eds.) WoLLIC 2016. LNCS, vol. 9803, pp. 432–445. Springer, Heidelberg (2016). https://doi.org/10.1007/978-3-662-52921-8_26

Zuber, R.: Reflexive and reciprocal determiners. In: Hansen, H.H., Murray, S.E., Sadrzadeh, M., Zeevat, H. (eds.) TbiLLC 2015. LNCS, vol. 10148, pp. 185–201. Springer, Heidelberg (2017a). https://doi.org/10.1007/978-3-662-54332-0_11

Zuber, R.: Set partitions and the meaning of the same. J. Logic Lang. Inform. **26**, 1–20 (2017b)

Author Index

Printed in the United States
By Bookmasters